# 3rd International Conference on Mechanical Metrology (CIMMEC 2014)



Journal of Physics: Conference Series Volume 648

Gramado, Brazil
14 – 16 October 2014

**Fejups;!**

**Spesjhp!Dptub.Gfmjy**

ISBN: 978-1-5108-1475-2
ISSN: 1742-6588

**Printed from e-media with permission by:**

Curran Associates, Inc.
57 Morehouse Lane
Red Hook, NY 12571

**Some format issues inherent in the e-media version may also appear in this print version.**



Printed by Curran Associates, Inc. (2016)



Institute of Physics
Dirac House, Temple Back
Bristol BS1 6BE UK

Phone: 44 1 17 929 7481
Fax: 44 1 17 920 0979

techtracking@iop.org

**Additional copies of this publication are available from:**

Curran Associates, Inc.
57 Morehouse Lane
Red Hook, NY 12571 USA
Phone: 845-758-0400
Fax: 845-758-2634
Email: curran@proceedings.com
Web: www.proceedings.com

# Table of contents

**Volume 648**

**3rd International Congress on Mechanical Metrology (CIMMEC2014)**
**14–16 October 2015, Gramado, Rio Grande do Sul, Brazil**

Journal of Physics: Conference Series **648** (2015) 011001 doi:10.1088/1742-6596/648/1/011001

**THE THIRD INTERNATIONAL CONGRESS ON MECHANICAL METROLOGY (CIMMEC 2014)**

From October 14th to 16th 2014, The Brazilian National Institute of Metrology, Quality, and Technology (Inmetro) and the Brazilian Society of Metrology (SBM) organized the 3rd International Congress on Mechanical Metrology (3rd CIMMEC). The 3rd CIMMEC was held in the city of Gramado, Rio Grande do Sul, Brazil. Anticipating the interest and enthusiasm of the technical–scientific community, the Organizing Institutions invite people and organizations to participate in this important congress, reiterating the commitment to organize an event according to highest international standards. This event has been conceived to integrate people and organizations from Brazil and abroad in the discussion of advanced themes in metrology.

Manufacturers and dealers of measuring equipment and standards, as well as of auxiliary accessories and bibliographic material, had the chance to promote their products and services in stands at the Fair, which has taken place alongside the Congress.

The 3rd CIMMEC consisted of five Keynote Speeches and 116 regular papers. Among the regular papers, the 25 most outstanding ones, comprising a high quality content on Mechanical Metrology, were selected to be published in this issue of *Journal of Physics: Conference Series*.

It is our great pleasure to present this volume of *Journal of Physics: Conference Series* to the scientific community to promote further research in Mechanical Metrology and related areas. We believe that this volume will be both an excellent source of scientific material in the fast evolving fields that were covered by CIMMEC 2014.

**PRESIDENT OF THE CONGRESS**

Humberto Siqueira Brandi
Brazilian National Institute of Metrology, Quality and Technology
hsbrandi@inmetro.gov.br

**VICE-PRESIDENTS OF THE CONGRESS**

José Carlos Valente de Oliveira
Brazilian National Institute of Metrology, Quality and Technology
jcoliveira@inmetro.gov.br

João Lerch
President of the Brazilian Society of Metrology (SBM)
secexec@metrologia.org.br

**CHAIR OF THE ORGANIZING COMMITTE**

Gelson Martins da Rocha
Brazilian National Institute of Metrology, Quality and Technology
gmrocha@inmetro.gov.br

**EDITOR-IN-CHIEF**

Rodrigo Costa-Félix
Brazilian National Institute of Metrology, Quality and Technology
rpfelix@inmetro.gov.br

Journal of Physics: Conference Series **648** (2015) 011001 doi:10.1088/1742-6596/648/1/011001

**Technical and Scientific Committee for CIMMEC 2014**

Adauto de Oliveira Barros Neto (Inmetro and SBM / Brazil)
Andy Knot (NPL/UK)
Armando Albertazzi (LABMETRO/UFSC/Brazil)
Dalni Malta do Espírito Santo Filho (Inmetro/Brazil)
Deise Borges (Inmetro/Brazil)
Etiene Benine (UFRGS/Brazil)
Flávio Salema (Inmetro/Brazil)
Gustavo Palmeira Ripper (Inmetro/Brazil)
João Antônio Pires Alves (Inmetro/Brazil)
Jorge C. Torres (CENAM/Mexico)
Kazuto Iwakita (IPT/Brazil)
Kelly da Silva B. Barbosa (Inmetro/Brazil)
Luis Omar Becerra (CENAM/Mexico)
Miguel Viliesid (CENAM/Mexico)
Paulo Roberto Guimarães Couto (Inmetro/Brazil)
Richard Davis (BIPM/France)
Romeu Daroda (Inmetro/Brazil)
Sérgio Pinheiro de Oliveira (Inmetro/Brazil)
Telmo Roberto Strohaecker (UFRGS/Brazil)
Thais Medeiros (Executive secretary of SBM/ Brazil)
Valter Aibe (Dinam/Inmetro/Brazil)
Victor Manuel Loayza Mendoza (Inmetro/Brazil)
Wellington Santos Barros (Inmetro/Brazil)

# Peer review statement

All papers published in this volume of *Journal of Physics: Conference Series* have been peer reviewed through processes administered by the proceedings Editors. Reviews were conducted by expert referees to the professional and scientific standards expected of a proceedings journal published by IOP Publishing.

# Assuring the quality of results of test hardness IRHD*: IPT's case study*

**T Yojo[1,2], M J A C Miranda[1,2], C B Oliveira[1,2] and C Matteucci[1,3]**

[1] IPT- Institute for Technological Research, Av. Prof. Almeida Prado, 532 – Cidade Universitária, CEP 05508-901, Sao Paulo, SP, Brazil
[2] LAMM -Trees, Woods and Furniture Laboratory
[3] GGQ – Quality Administration Management

E-mail: yojos@ipt.br, mmiranda@ipt.br, claudiob@ipt.br, cyntiamt@ipt.br

**Abstract.** This paper presents the experience of the Laboratory Trees, Woods and Furniture - LAMM in calibrating the durometer IRHD and assuring the quality of its test results, since there are no Certified Reference Material and laboratory in the Brazilian Calibration Network that can calibrate the equipment. To solve this problem, the IRHD hardness (N method) was quantified in three ways: a) by measuring the modulus of elasticity of the material, b) by measuring the depth the sphere entered the material and c) the durometer's direct reading. With the IRHD hardness measured by accepted international standards techniques, it was possible to evaluate the accuracy of the test results that assured the calibration of the equipment.

## 1. Introduction

The devices and materials involved in the characterization tests and performance evaluation of melamine laminated flooring and furniture, require the use of polymeric materials with specified hardness values. Therefore, it is necessary to know these values in order to check for their compliance with the requirements of the related standards.

To perform this test the Laboratory Trees, Wood and Furniture - LAMM from the Technological Research Institute of São Paulo State S.A - IPT, acquired an IRHD hardness tester in the range 30 IRHD to 100 IRHD. After purchasing the equipment, the LAMM faced the problem of ensuring the metrological reliability, since there is no laboratory in the Brazilian Calibration Network (RBC) that could calibrate the equipment.

In addition, there isn't a producer of certified reference materials in Brazil besides, the purchase abroad does not guarantee prompt delivery. As a result, to solve the calibration problem of the equipment and to ensure the quality of the test results, the LAMM prepared samples of a selected polymeric material and subjected them to tests from the following accepted international standards: ISO 48: 2010 [1], ISO 18898: 2012 [2] and ISO 7743: 2011 [3], applicable to different polymeric materials. All tests were done under controlled conditions of temperature and relative humidity, according to the criteria established at ABNT NBR 13966: 2008 standards [4] and ABNT NBR 14833-1: 2009 [5].

With the data collected from the determination of the modulus of elasticity of the material, the measurement of the depth of indentation of the sphere and the direct reading on the tester, the results were compared via a statistical analysis [6,7] in order to check for the calibration of the equipment and to verify the quality assurance of the results [8].

## 2. Objective

This article aims to show the experience of the LAMM in the calibration procedure of an IRHD tester and to demonstrate the quality assurance of the results of an IRHD hardness test.

## 3. Methods

From a piece of polymeric material six circular shaped samples were cut with a diameter of 29.0 mm and thickness of 12.5 mm, they were conditioned at a relative air humidity of (50 ± 5) % and a temperature of (23 ± 2)° C for a period of 7 days.

### *3.1. Definition of hardness IRHD*

According to ISO 48: 2010, hardness IRHD by method N is a quantity defined as a function of the depth of indentation, expressed as (1/100) mm, of a steel sphere with a diameter of (2.50 ± 0.01) mm, on a polymeric material initially subjected to a force of (0.30 ± 0.02) N, under the action of an additional force of (5.40 ± 0.01) N. The relationship between the depth of indentation of the sphere 'D' in (1/100) mm and its hardness, is the IRHD shown in table 3 - Conversion of *D* values of IRHD for the N method, from the standard ISO 48 [1].

### *3.2. Weight and diameter of the sphere tester*

The diameter of the sphere was measured and the weight of the device was determined in order to verify whether they met the requirements of the standard ISO 48: 2010 [1].

### *3.3 Calibration via modulus of elasticity*

In this IRHD hardness calibration method it was used Equation 1 from ISO 48: 2010 [1], which calculates the indentation depth of the sphere and allows to express the value of IRHD hardness with the help of table 3 - Conversion of *D* values of IRHD for method N of standard ISO 48 [1], forementioned.

$$D = 61.5\ R^{-0.48}\left[\left(\frac{F_{in}}{E}\right)^{0.74} - \left(\frac{F_C}{E}\right)^{0.74}\right] \qquad (1)$$

where:

$D$ = the differential indentation (1/100) mm;
$R$ = the radius of the ball (mm);
$F_{in}$ = the total indenting force (N);
$F_c$ = the contact force (N);
$E$ = specimen's modulus of elasticity (MPa).

As previously mentioned, the IRHD hardness is attributed to the indentation of a sphere of 2.50 mm in diameter, under an contact force of 0.30 N in addition to a force of 5.40 N, resulting in the total indenting force of 5.70 N.

The determination of the modulus of elasticity (E) was in accordance with ISO 7743: 2011 [3], with a deformation up to 5 %.

Having obtained the value of E, equation (1) was used to estimate D namely, the relative indentation. Referring to the value of the D from Table 3 - Conversion of D values of IRHD for method N from standard ISO 48, the corresponding value of IRHD hardness was obtained.

### *3.4 Calibration via the depth of* indentation *of the sphere*

This test was based on the criteria established in ISO 48: 2010 [1] and ISO 18898: 2012 [2].

The device used for measuring the depth of indentation of the sphere consisted of a photographic camera with a calibrated scale and with standard of reference, attached to a digital microscope Dino-Lite Pro, model AM413T PRO. The depth of indentation was determined by taking two photos with the still camera and measuring reference points before and after the application of the force.

With the values of the depth of indentation of the sphere, the IRHD hardness was obtained, according to the standard ISO 18898: 2012 [2].

*3.5 Determination of hardness IRHD N with tester*

The determination IRHD hardness was performed directly, according to the criteria of ISO 48: 2010 [1], using the IRHD hardness tester N.

*3.6 Precision Experiment*

To determine the precision under the condition of repeatability, according to ISO 5725-4: 1994 [6], IRHD hardness tests were carried out on eight samples with three replicates, using the hardness tester IRHD N.

## 4. Results

*4.1. Weight and diameter of the sphere tester*

The statistics of the measurements on ball diameter and weight of the device, are presented in table 1.

**Table 1.** Statistical data obtained from the sphere diameter and weight of the device.

| Statistics [a] | Diameter (mm) | Weight (N) |
|---|---|---|
| Average | 2.497 | 5.396 |
| Standard deviation | 0.00063 | 0.00005 |
| Reference value (ISO 48) | 2.50 | 5.40 |
| Error = average - reference | - 0.003 | - 0.004 |
| Tolerance (ISO 48) | ± 0.01 | ± 0.01 |

[a]Results of 10 measurements.

*4.2. Hardness IRHD N*

The IRHD N hardness results calculated from the modulus of elasticity and the depth of indentation of the sphere and those obtained directly from the durometer readings according to the methods presented, respectively, in items 3.3, 3.4 and 3.5 are presented in table 2.

Journal of Physics: Conference Series **648** (2015) 012001 doi:10.1088/1742-6596/648/1/012001

**Table 2.** Results of IRHD hardness calculated from the modulus of elasticity, the depth of indentation of the sphere and those obtained directly from a tester.

| Specimens | Modulus of elasticity | | | Measurement of the depth of indentation of the sphere | | Durometer reading |
|---|---|---|---|---|---|---|
| | $E$ [a] | $D$ [b] | Hardness IRHD N | $D$ [c] | Hardness IRHD N | Hardness IRHD N |
| 1 | 4.64 | 57 | 67.1 | 68 | 61.5 | 62.2 |
| 2 | 4.34 | 60 | 65.5 | 68 | 61.5 | 62.0 |
| 3 | 4.54 | 58 | 66.6 | 72 | 59.7 | 62.0 |
| 4 | 4.57 | 58 | 66.6 | 72 | 59.7 | 62.0 |
| 5 | 7.10 | 42 | 75.8 | 48 | 72.2 | 71.1 |
| 6 | 4.77 | 56 | 67.6 | 68 | 61.5 | 61.4 |
| Average | 4.99 | 55 | 68.2 | 66 | 62.7 | 63.4 |
| Standard deviation | 1.04 | 6.7 | 3.0 | 9.0 | 4.7 | 3.8 |

Legend:
[a] $E$ = modulus of elasticity of the specimen (MPa).
[b] $D$ = differential indentation calculated from in the equation using the modulus of elasticity (1/100) mm.
[c] $D$ = differential indentation obtained from direct measurement (1/100) mm.

The results of the statistical analysis of the IRHD hardness obtained via the methods of the modulus of elasticity and the measurement of the depth of indentation of the sphere when compared with those obtained directly from the tester, are shown in table 3.

**Table 3.** Results of the IRHD hardness obtained via the methods of the modulus of elasticity and the measurement the of the depth of indentation of the sphere.

| Statistics [a] | Method: Module of elasticity | Method: Direct measurement of the depth |
|---|---|---|
| Mean difference | 4.8 | -0.8 |
| Standard deviation | 0.864 | 1.343 |
| t calculated | 13.462 | -1.399 |
| Degrees of freedom | 5 | 5 |
| Results | Significant difference | Non significant difference |

[a] t critical = 2.015
Difference = readings from the method - readings from the tester

Figure 1 shows the comparison among the hardness results estimated by the proposed methods and their respective regression equations.

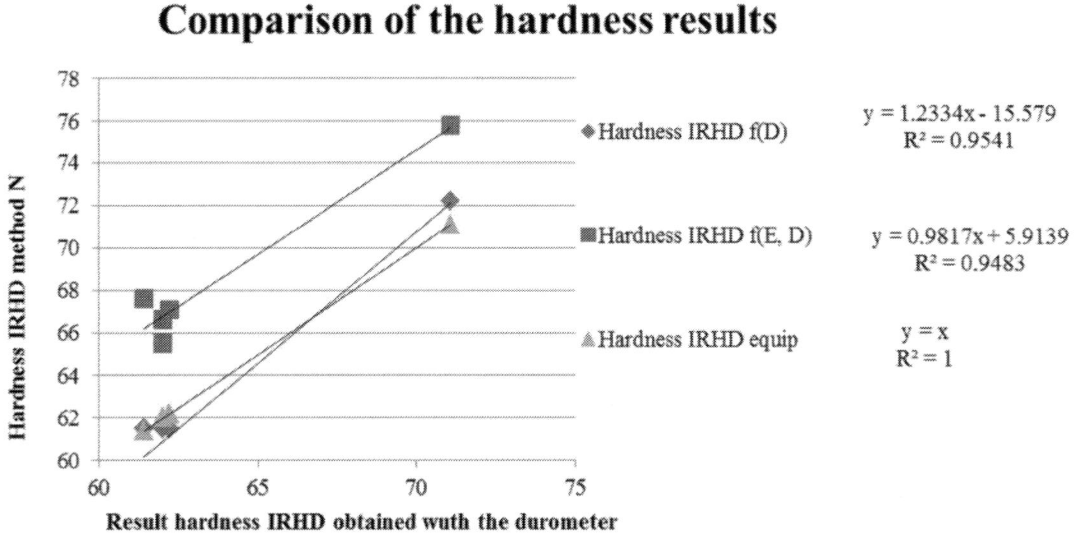


**Figure 1.** Results of the hardness IRHD N estimated and the respective regression equations.

*4.3. Experiment of the Precision of the durometer IRHD N*

The results of the test for determining hardness IRHD obtained with the IRHD hardness tester, to evaluate the repeatability are presented in table 4.

**Table 4.** Results of the test for determining hardness IRHD obtained with the IRHD hardness tester, to evaluate the repeatability.

| Specimens | IRHD | | | | |
|---|---|---|---|---|---|
| | 1 | 2 | 3 | Averages | Variance $Si^2$ [a] |
| 1 | 61.4 | 61.4 | 61.0 | 61.3 | 0.05 |
| 2 | 60.5 | 60.7 | 60.6 | 60.6 | 0.01 |
| 3 | 61.2 | 61.4 | 61.3 | 61.3 | 0.01 |
| 4 | 60.7 | 61.1 | 60.8 | 60.9 | 0.04 |
| 5 | 61.0 | 61.3 | 60.6 | 61.0 | 0.12 |
| 6 | 61.3 | 61.3 | 61.7 | 61.4 | 0.05 |
| 7 | 60.7 | 61.0 | 61.3 | 61.0 | 0.09 |
| 8 | 61.1 | 61.4 | 61.1 | 61.2 | 0.03 |
| $Sr^2$ [b] | | | | | 0.05 |
| repeatability | | | | | 0.22 |

[a] Compliance to Equation 9, standard ISO 5725-4 [6].
[b] Compliance 8, standard ISO 5725-4 [6].

The repeatability obtained for the durometer of IRHD was of 0.22 hardness IRHD unit.

Journal of Physics: Conference Series **648** (2015) 012001 doi:10.1088/1742-6596/648/1/012001

## 5. Discussion

It has been demonstrated the suitability of the tester by verifying the compliance of the results obtained by the weight (5.40 N) and the sphere diameter (2.50 mm), shown in table 1, to the limits specified in ISO 48: 2010 [1].

The manufacturer of the tester sent, at the time of acquisition of the equipment, a calibration certificate n° 10620, of 25/08/2011, issued by the Deutscher Kalibrierdienst DKD, calibration label: DKD-K-16501.

In this certificate the uncertainty in the detected sag was of 0.3 hundredths of a millimeter, which corresponds to about 0.15 to 0.18 unit the IRHD hardness, based on table 3 of the ISO 48 standard for IRHD hardness, in the range of 53-66.

In table 2 it can be seen a non-significant difference between the values determined of the hardness IRHD with a durometer (63.4) and the calibration by measuring the depth of indentation of the sphere (62.7).

This non-significant difference can be explained by the fact that between the two measurements, according to the method of determination, the only variable that was changed was the way of measuring the indentation of the sphere.

In the case of calibration via modulus of elasticity of the material, see table 2, the difference was significant, because this method tends to estimate higher hardness values them the ones obtained with the tester.

On the other hand estimation of the modulus of elasticity (E) of the polymeric material is influenced by many variables that contribute to the uncertainty of the result such as: force, deformation, specimen dimensions, contact surface lubrication and positioning of the specimen.

Regarding to the results of the precision experiment, the repeatability of 0.22 in the IRHD hardness obtained with the tester, is compatible with the value expressed in the Calibration Certificate of the equipment.

## 6. Conclusion

From the results obtained it can be concluded that:

a. The determined values of the sphere diameter and the load of the device IRHD hardness tester are in accordance with ISO 48 [1], Method N.
b. The estimated values of IRHD hardness obtained via the modulus of elasticity method, showed significant difference from those obtained with IRHD hardness tester.
c. The results of hardness obtained with the IRHD hardness tester were not significant from those estimated via the measurement of the depth of indentation of the sphere method.
d. The method of measuring the depth of indentation [1] to calibrate IRHD tester results, showed to be adequate, since it allowed quantifying the displacements at different loading points by optical means without physical contact.
e. It was possible to demonstrate the quality assurance of the results of the IRHD hardness test.
f. New test procedures can be developed and offered internationally to the technological environment in order to improve the services to the users of standardized materials.

## 7. References

[1] International Standard Organization 2010 ISO48 – *Rubber, vulcanized or thermoplastic – Determination of hardness (hardness between 10 IRHD and 100 IRHD)* (Berlin, ISO) p 23

[2] International Standard Organization 2012 ISO18898 - *Rubber – Calibration and verification of hardness testers* (Geneva, ISO) p 18

[3] International Standard Organization 2011 ISO7743 – *Rubber, vulcanized or thermoplastic – Determination of compression stress-strain properties* (Geneva, ISO) p 22

[4] Associação Brasileira de Normas Técnicas 2008 ABNT NBR 13966 – *Móveis para escritório – Mesas – Classificação e características físicas dimensionais e requisitos e métodos de ensaio* (Rio de Janeiro, ABNT) p 19

[5] Associação Brasileira de Normas Técnicas 2009 ABNT NBR 14833-1 – *Revestimento de pisos laminados melamínicos de alta resistência- Parte 1: Requisitos, características, classes e método de ensaio.* (Rio de Janeiro, ABNT) p 42

[6] International Standard Organization 1994 ISO 5725-4 –*Accuracy (trueness and precision) of measurement methods and results – Part 4: Basic methods for the determination of the trueness of a standard measurement method* (Geneva, ISO) p 23

[7] Costa Neto P L O 1977 Estatística, (São Paulo, Edgard Blucher) p 264

[8] Associação Brasileira de Normas Técnicas 2005 ABNT NBR 17025 – *Requisitos gerais para a competência de laboratórios de ensaio e calibração.* (Rio de Janeiro, ABNT) p 31.

# Investigation of the causes for the measurement differences between the natural gas measurement systems of a gas distribution company and of a consumer

**Kazuto Kawakita[1] and Rubens Silva Telles[2]**

[1, 2] IPT-Instituto de Pesquisas Tecnológicas
Centro de Metrologia Mecânica, Elétrica e de Fluidos-CTMetro
Av. Prof. Almeida Prado, 532 - Butantã
05508-901 São Paulo, Brasil

kawakita@ipt.br

**Abstract**. This paper presents the methodology used, the checked items and the results obtained in an investigation carried out to identify the causes that generated substantial differences between the natural gas volumes measured by two measuring systems, one from a gas distribution company and another from its industrial consumer, both installed in series in a gas pipeline. The investigation showed that the measurement of the volume of gas performed by the gas distributor metering system was influenced by an erroneous measurement of the gas temperature since it was affected by a complex thermodynamic process involving cooling by the Joule-Thomson effect caused by a pressure reducing valve and heating by heat exchange through the pipe walls.

## 1. Introduction

From an economic point of view, one of the most important activities in the natural gas industry is related to the measurement of the volume of gas produced, processed, transported and commercialized. The reliability in achieving this metrological activity is essential for all companies operating along the entire value chain of this strategic industry for many countries. To obtain this metrological reliability it is necessary to use gas flow measurement systems in most stages of the chain, from production of the gas onshore or offshore or in its importation as a liquefied natural gas, passing through the treatment units, transportation through the long pipelines at last coming to the distribution to final consumers.

Natural gas is a high-value commodity, so it is typical that the parties involved install in all custody transfer stations or delivery points to a major consumer, two independent measuring systems, one operated by the gas supplier and the other by the receiver. Usually, the billing of gas volumes traded is based on the measurement performed by the measurement system from the gas deliverer, although this measurement is continuously being monitored by the measuring system of the receiver.

Nevertheless the sophisticated measurement technology used nowadays, the occurrence of differences between the gas volumes measured by the two measuring systems is quite common, and occasionally these differences go far beyond the limits agreed in the contracts signed by the parties.

The main reason for that is the inherent difficulty in measuring the flow rate of fluids, which is a dynamic phenomenon, dependent on the characteristics of the fluid measured and which can be affected by the different influence quantities impacting the metrological process.

## 2. Objective

This paper presents the methodology followed, the checked items and the results obtained in an investigation carried out on two natural gas measurement systems, one operated by a natural gas distribution company and the other by an industrial consumer, and that aimed to determine the reasons that caused significant differences between the gas volumes measured by the parties.

Installed in series, the measuring system of the gas distributor used a turbine meter while the industry measuring system was based on a cone type differential pressure flowmeter.

For reasons of secrecy and confidentiality, the names of the companies will be omitted in this work.

## 3. Characteristics of the measurement systems

The gas measurement system operated by the gas distributor was based on a turbine meter, assembled on a 10 inches diameter horizontal meter tube, also composed of a straight pipe length upstream of the meter, a downstream pipe section, and a flow straightener. The measurement system included a flow computer with integrated sensors/transmitters for measuring the gauge pressure and temperature of the gas. This assembly was mounted as a measurement skid.

Figure 1, below, illustrates schematically the flow measuring system of the gas distributor and its main components.

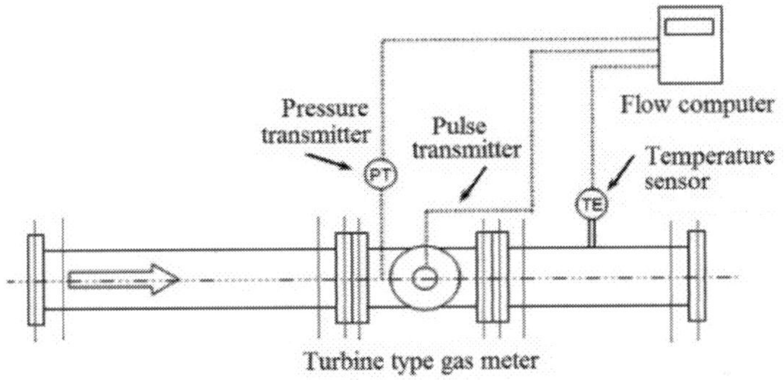


**Figure 1.** Measuring system of the gas distribution company based on a turbine type gas meter.

In its turn, the natural gas flow measuring system installed in the industrial plant gate was composed of a cone type flow meter, assembled on a 10 inches horizontal pipe, completed by straight pipe sections upstream and downstream of the meter. This measurement system included also a flow computer associated to an integrated sensor/transmitter for measuring the gauge pressure, differential pressure and temperature of the gas.

Figure 2, below, illustrates schematically the flow measurement system of the industry and its main components.

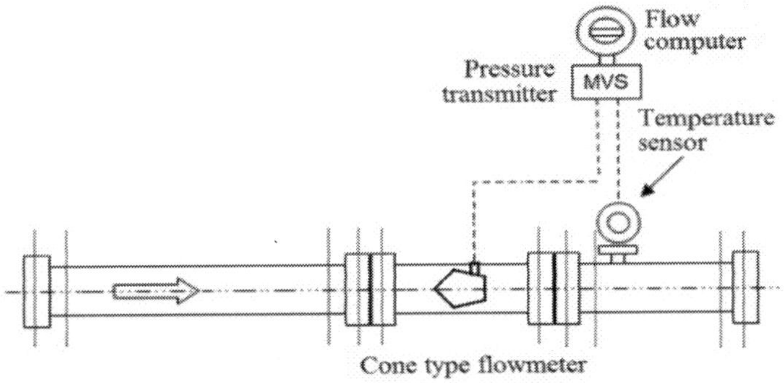


**Figure 2.** Gas measurement system of the industry based on a cone type flowmeter.

Journal of Physics: Conference Series **648** (2015) 012002 doi:10.1088/1742-6596/648/1/012002

## 4. Methodology

From the metrological point of view, the inspections of the gas flow measurement systems were carried out in order to assess the reliability of the natural gas volumes measured by the systems in accordance with the requirements and procedures set out in the reference standards.

The reference documents used in this study were the AGA Report No. 7 [1], the AGA Report No. 8 [2], the GUM guide [3] and the ISO 5168 standard [4].

Considering the case of two natural gas measurement systems, one based on a cone type differential pressure meter and the other on a turbine meter, the inspection was carried out by checking the construction, installation and operation parameters of the components of each system.

### *4.1. Inspection of the primary elements*

The gas distributor metering system was visually inspected in order to check if the meter tube upstream and downstream straight pipe sections, and also the turbine meter body itself were assembled according to the requirements defined in the AGA Report No. 7 standard.

Although there is no technical standard to be followed in case of installation and use of cone type meters, a visual inspection of the meter tube of the industry measurement system was also carried out in order to analyze the compliance with the manufacturer's installation instructions for this type of meter.

Additionally, the inspection evaluated if the installation positions of the pressure and temperature sensors/ transmitters were in compliance to the assembling requirements of the reference standard for a turbine type measuring system.

### *4.2. Calibration of the secondary elements*

As a second step of the inspection, calibrations of the gauge pressure, differential pressure and temperature sensors and transmitters of the industry and of the gas distributor measurement systems were carried out using standards calibrated and certified by ISO 5167 accredited laboratories.

Calibrations of the sensors and transmitters of both measurement systems were performed in closed loop, i.e., applying input signals in the pressure and temperature sensors and then reading the parameters values directly in the flow computers.

### *4.3. Inspection of the flow computers*

Inspection of the two flow computers intended to verify the various parameters entered in their configuration, such as measurement units, base pressure and temperature conditions, local barometric pressure, set points of the electronic transmitters for gauge pressure, differential pressure and temperature quantities, composition and properties of the natural gas, technical reference standards, among others.

Additionally, the gas volumes calculated by the flow computers were checked in order to validate the calculation algorithms used by these equipment.

### *4.4. Assessment of the measurement uncertainty*

As a final part of the inspection process, the measurement uncertainties associated to the volumes of gas measured by both systems were estimated for real operating conditions. For this, the contributions of the uncertainties of each factor (gauge pressure, differential pressure, temperature, gas composition, meter factor, among others) we considered in the assessment of the overall uncertainty.

## 5. Results

### *5.1. Results of visual inspection of the primary elements*

It was observed that in the gas distributor measurement system, the upstream and downstream straight pipe lengths complied with the installation requirements of the reference standard.

Likewise, the installation positions of the pressure taps and temperature thermowell met the requirements of the standard.

However, as can be seen in figure 3, the measurement skid of the gas distributor included valves, reducers and 90° bends in perpendicular planes at the inlet of the straight pipe section upstream of the turbine meter. This set up can induce disturbances in the flow and which could cause measurement errors according to item 7.1.7 of the AGA Report No. 7 standard.

The most common disturbances in these situations are the asymmetry in the gas velocity profile and the helical flow of the gas, frequently known as swirl. It is well known that the turbine flowmeter performance is affected, to a greater or lesser extent, by these effects, but the exact amount is difficult to quantify, requiring for this experimental tests on laboratory test benches.

**Figure 3.** The gas distributor measurement system with turbine meter.

The physical inspection of the measurement system of the industry showed no abnormality of installation since it included long straight pipe sections upstream and downstream of the cone type meter, complying with the manufacturer's instructions for its installation.

**Figure 4.** Measuring system of the industry with a cone type flowmeter.

*5.2. Calibration results of the secondary elements*

The gauge pressure, differential pressure and temperature sensors and transmitters were calibrated in closed loop with the flow computer.

It was detected during the preliminary calibration of the instruments that the gauge pressure transmitter installed in measurement system of the gas distributor showed an indication error of -1.8 % of the span. Under request of the gas distributor agent, the instrument was calibrated, but not adjusted, i.e., it was kept under the same conditions found during the inspection mainly for preserving its status for necessary corrections.

### *5.3. Flow computer inspection results*

The parameters of the setup menus and the gas composition inserted into the two flow computers were verified.

During the inspection of the flow computer of the industry measurement system, it was detected that the natural gas composition parameterized in it was different from that parameterized in the flow computer of the gas distributor measuring system. Under request of the industry agent the gas composition configuration was update in its flow computer.

It was checked that the flow computer of the gas distributor measurement system performed the calculations of gas volumes correctly. However, the flow computer of the industry's measurement system showed deviations of up to about 0.55 % compared to the volumes calculated by checking program used.

Results of the verification of the flow computers calculation algorithms of the gas measurement systems of industry and gas distributor are presented respectively in tables 1 and 2 below.

**Table 1.** Results of the verification of the industry flow computer calculation algorithm.

| | | **Differential pressure [mmH2O]** | | | **Deviation [%]** | | |
|---|---|---|---|---|---|---|---|
| | | **250** | **300** | **2 275.3** | | | |
| **Gauge pressure [kPa]** | | **200** | **300** | **400** | | | |
| **Temperature [°C]** | **0** | 5 440.80 | 6 933.63 | 21 013.35 | **0.342** | **0.058** | **0.335** |
| | | **5 459.47** | **6 937.63** | **20 943.30** | | | |
| | **10** | 5 346.15 | 6 811.78 | 20 640.27 | **0.218** | **0.049** | **0.442** |
| | | **5 357.80** | **6 808.45** | **20 549.50** | | | |
| | **20** | 5 256.76 | 6 696.83 | 20 288.76 | **0.095** | **0.155** | **0.545** |
| | | **5 261.78** | **6 686.43** | **20 178.10** | | | |

**Table 2.** Results of the verification of the gas distributor flow computer calculation algorithm.

| | | **Gauge pressure [kPa]** | | | **Deviation [%]** | | |
|---|---|---|---|---|---|---|---|
| | | **400** | **400** | **600** | | | |
| **Uncorrected flowrate** | | **120** | **160** | **160** | | | |
| **Temperature [°C]** | **0** | 6 261,773 | 8 349,031 | 11 855,494 | **0,014** | **0,014** | **0,016** |
| | | **6 262,620** | **8 350,160** | **11 857,370** | | | |
| | **10** | 6 030,125 | 8 040,167 | 11 408,390 | **0,014** | **0,014** | **0,016** |
| | | **6 030,940** | **8 041,250** | **11 410,200** | | | |
| | **20** | 5 815,612 | 7 754,149 | 10 995,448 | **0,014** | **0,014** | **0,016** |
| | | **5 816,400** | **7 755,200** | **10 997,190** | | | |

### *5.4. Results of the measurement uncertainty assessment*

Based on the information and data collected during the inspection of the metering systems, it was possible to estimate the uncertainty associated with the measured volumes of natural gas, but only for the measurement system of the industry. This measurement system, operating under normal conditions of process parameters, is capable of obtaining uncertainties of approximately 1.6 % of the measured volume, considering a confidence level of approximately 95 %.

This measurement uncertainty assessment was developed based on the methodology presented in the GUM 2008 - *Evaluation of measurement data – Guide to the expression of uncertainty in measurement* [3] and ISO 5168: 2005 [4]. In this uncertainty estimation, besides the uncertainty contributions arising from the calibration of the cone type meter, other sources of uncertainty were considered such as the uncertainties associated with the calibration of the secondary elements, the natural gas expansion and compressibility factors and the differences detected in the flow computer calculation algorithm.

On the other hand, the uncertainty estimation associated to the volume of gas measured by the gas distributor metering system was not performed because it was not possible to estimate the uncertainty contributions in the measurement of the gas temperature in the meter due to the conditions found at the site. These restrictions are mainly due to the unfeasibility to assess the resulting uncertainties from:

(a) the effects caused by the expansion of gas due to the abrupt reduction of its pressure in the pressure reducing valve;
(b) the rates of gas heat exchange with the walls of the pipe, which depend on the environmental conditions at the moment (night or day, winter or summer), and
(c) the effects of disturbances in the flow at the turbine meter inlet, due to the physical configuration of the pipes upstream of the meter in the measuring skid.

Table 3 shows the results of the measurement uncertainty assessment performed for the gas measurement system of the industry.

| Input quantity $X_i$ | Value $x_i$ | Std. unc. $u(x_i)$ | Sensibility coefficient | Uncertainty | Contribution (%) |
|---|---|---|---|---|---|
| Discharge coefficient | 0.82920 | 0.0014 | 1.934E+00 | 0.002681 | 4.7 % |
| Apr. Velocity factor | 1.51613 | 6.02566E-05 | 1.058E+00 | 0.000064 | 0.0 % |
| Expansion coefficient | 0.9968 | 8.04002E-05 | 1.609E+00 | 0.000129 | 0.0 % |
| Geometric dimension | 0.22144 m | 0.000005 m | 1.449E+01 | 0.000072 | 0.0 % |
| Differential pressure | 2921 Pa | 42.315 Pa | 2.746E-04 | 0.011619 | 89.0 % |
| Gas density | 3.739 kg/m³ | 0.0144 kg/m³ | 2.145E-01 | 0.003080 | 6.3 % |
| Flow computer | 1.604 kg/m³ | 0.0044 kg/m³ | 1.000E+00 | 0.004374 | 12.6 % |
| Mass flowrate $Q_m$ | **7.13 kg/s** | **Comb. std. unc. $u_c$ ($Q_m$) =** | | **0.0123 kg/s** | **100 %** |
| **Coverage factor $k$ = 2,00** | | | | | |
| **Expanded uncertainty $U$ ($Q_m$) = 0.02464 kg/s** | | | | | |

| Input quantity $X_i$ | Value $x_i$ | Std. unc. $u(x_i)$ | Sensibility coefficient | Uncertainty | Contribution (%) |
|---|---|---|---|---|---|
| Mass flowrate $Q_m$ | 1.604 kg/s | 0.0123 | 1.2988 | 0.015997 | 86.9 % |
| Gas density (base cond.) | 0.770 kg/m³ | 0.0023 | -2.7058 | -0.006114 | 13.1 % |
| Volum. flowrate $Q_{vb}$ | **180 000 m³/day** | **Comb. std. unc. $u_c$ ($Q_{vb}$) =** | | **0.01716 m³/s** | **100 %** |
| **Coverage factor $k$ = 2.00** | | | | | |
| **Expanded uncertainty $U$ ($Q_{vb}$) = 2966 m³/day (1.6 %)** | | | | | |

The declared expanded uncertainty is based on a combined standard uncertainty multiplied by a coverage factor $k$ = 2.00 under a confidence level of approximately 95 %.

Obs.: $Q_{vb}$ is the volumetric flowrate of gas at base conditions of 20 °C & 101.325 kPa.

**Table 3.** Measurement uncertainty estimation for the measurement system of the industry.

As can be seen from the calculation, the industry measurement system enabled gas volume measurements with an associated uncertainty of approximately 1.6 %.

### *5.5. Measuring differences between the two measurement systems*

The hourly rate volumes of gas measured by the two measurement systems were converted to the base conditions of 20 °C and 101.325 kPa, typically used in the gas industry.

The flow computer of the measurement system operated by the gas distribution company calculated and stored the average gas pressure values, average temperature, uncorrected volume and corrected volume computed in a hourly basis.

The multivariable transmitter, which had a unit to totalize the gas volumes measured, sent these values to a supervisory system in the gas distributor's control room. This supervisory system also received the values totalized by the measurement system of the gas distributor.

During the inspection period, data of gas hourly volume measurements for different days were collected, from which significant differences between the two volumes computed by the two measurement systems were identified. In fact, along the period analyzed, differences of up to 5 % were observed between the gas hourly volumes totalized by these two measurement systems.

Table 4 shows the gas hourly volumes measured by the industry measurement system and the gas distributor system and the percentage differences for a particular day.

**Table 4.** Hourly volumes of gas measured by the two measurement system and the differences.

| Hour | Hourly volume of gas measured by the industry measuring system [m³] | Hourly volume of gas measured by the gas distributor measuring system [m³] | Percentage difference [%] |
|---|---|---|---|
| 00:00 | 6 709.11 | 6 966.58 | 3.84 |
| 01:00 | 6 930.47 | 7 173.85 | 3.51 |
| 02:00 | 6 672.09 | 6 928.17 | 3.84 |
| 03:00 | 6 463.44 | 6 726.61 | 4.07 |
| 04:00 | 6 621.70 | 6 872.58 | 3.79 |
| 05:00 | 6 739.50 | 6 994.28 | 3.78 |
| 06:00 | 6 655.45 | 6 943.12 | 4.32 |
| 07:00 | 6 797.46 | 7 043.78 | 3.62 |
| 08:00 | 6 706.10 | 6 976.76 | 4.04 |
| 09:00 | 6 519.92 | 6 631.93 | 1.72 |
| 10:00 | 6 716.59 | 6 739.39 | 0.34 |
| 11:00 | 6 604.67 | 6 558.60 | -0.70 |
| 12:00 | 6 741.12 | 6 716.46 | -0.37 |
| 13:00 | 6 709.22 | 6 621.88 | -1.30 |
| 14:00 | 6 827.17 | 6 711.34 | -1.70 |
| 15:00 | 6 690.56 | 6 580.43 | -1.65 |
| 16:00 | 6 589.48 | 6 480.57 | -1.65 |
| 17:00 | 6 444.30 | 6 342.79 | -1.58 |
| 18:00 | 6 548.35 | 6475.01 | -1.12 |
| 19:00 | 6 551.96 | 6 602.19 | 0.77 |
| 20:00 | 6 353.72 | 6 499.35 | 2.29 |
| 21:00 | 6 499.98 | 6 674.18 | 2.68 |
| 22:00 | 6 639.01 | 6 840.80 | 3.04 |
| 23:00 | 6 576.51 | 6 817.78 | 3.67 |
| 00:00 | 6 711.96 | 6 924.58 | 3.17 |

Figure 5 shows the results of hourly volumes of gas at reference conditions measured by the gas measurement systems of the industry and the gas distributor, and also the temperature and pressure of the gas measured by the measuring system of the gas distributor along the same day.

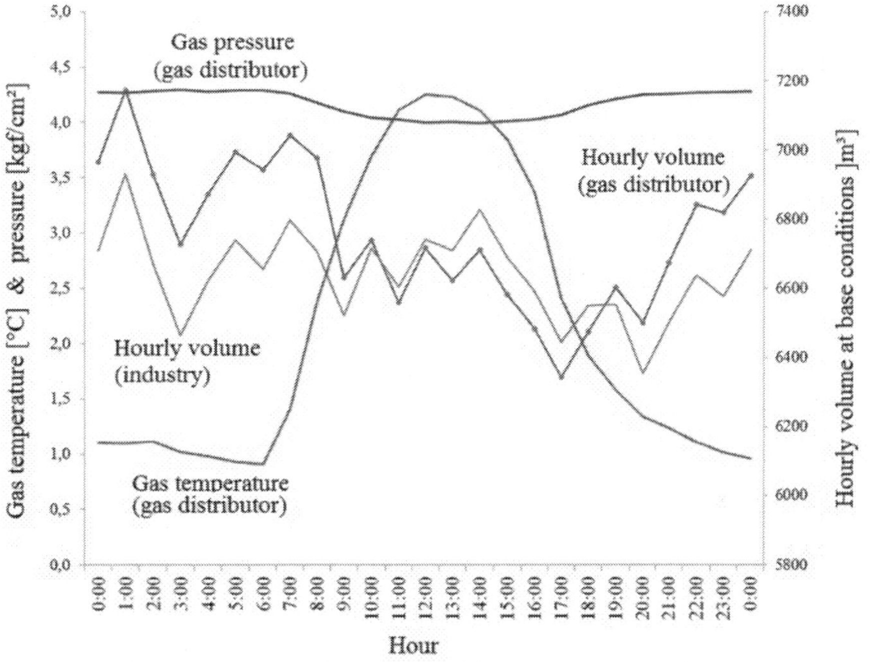


**Figure 5.** Hourly volumes of gas at reference conditions measured by the two gas measurement systems and the temperature and pressure of the gas measured by the measuring system of the gas distributor.

## 6. Analysis

An important fact observed during the inspection of the gas distributor measurement system is that the metering skid in question included a pressure reducing valve that operated reducing the pressure of the gas received via the pipeline from 35 kgf/cm² to 4 kgf/cm² in one single step. The Joule-Thomson cooling effect due to the abrupt expansion of the gas caused a great reduction in its temperature, as evidenced by the high rate of water vapor from ambient air condensed on the pipe walls downstream the pressure reducing valve.

After this cooling process downstream the valve, the gas flowed along the pipe exchanging heat with the pipe walls, and getting warm gradually reaching the turbine flow meter at a temperature not yet fully stabilized. This thermodynamic process includes internal temperature gradients in each cross section, as illustrated in the computational simulation of the gas temperature levels shown in figure 6.

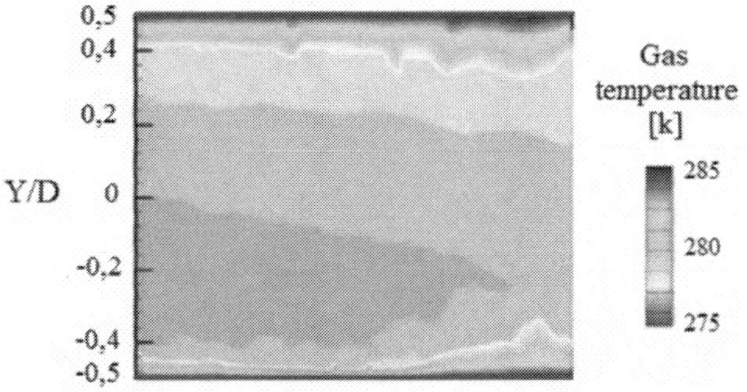


**Figure 6.** Gas temperature gradients inside the pipe.

This rate of heat transfer depends on environmental conditions (air temperature, wind, sunlight) at the moment, so that the measurement result is also influenced by these factors. Thus, the temperature measured by the sensor installed after the turbine may not exactly represent the average gas temperature in the flow. Again, the estimation of this effect influenced by multiple parameters is by no means a simple phenomenon and would require for that extensive simulation tests in a laboratory test bench.

## 7. Conclusion

The study made it possible to observe that the behavior of the measured parameters and influence quantities shown in table 3 and figure 5 repeated in a similar way along other days. These results indicated that the measurement of the volume of gas carried out by the gas distributor metering system was influenced by an erroneous measurement of the gas temperature. This fact is evidenced in figure 5 by the line of the gas temperature variation, where it is observed a sharp increase of this temperature in the early hours of the day and also a decrease in it during the afternoon and a relatively stable situation overnight, indicating a correlation of the measured temperature of the gas with the ambient temperature.

Thus, a hypothesis that could explain this behavior of the gas measurement system of the gas distributor is that the gas cooling effect due to its expansion in the pressure reducing valve is more efficiently mitigated by the heat exchange between the gas and the pipe during the day than during the night. Thus, an underestimation of the gas temperature overnight tends to result in a calculation of a larger density for the natural gas and, consequently, an overestimation of the gas volume totaled in this period. This fact caused a larger difference between the volumes measured by the two measurement systems during the night period.

In view of the issue addressed in this article, a warning should be made to the responsibles for the measurement management in the gas industry since there are many natural gas measurement systems installed excessively close to pressure reducing valves and that may be suffering from this same problem reported herein.

## References

[1] AGA Report No. 7 Measurement of Gas by Turbine Meters, ed. 2006.
[2] AGA Report #8 Compressibility Factors of Natural Gas and Other Related Hidrocarbon Gases 1994-1999.
[3] GUM 2008 - Evaluation of measurement data – Guide to the expression of uncertainty in measurement.
[4] ISO 5168 Measurement of fluid flow – Evaluation of uncertainties, 2nd Edition, 2005.

# Magnetic system for the quality control of specimens for Charpy impact test

**R V Martin**[1], **M A P Castanho**[2]
[1, 2] Instituto de Pesquisas Tecnológicas do Estado de São Paulo - IPT, Cidade Universitária, São Paulo, SP, 05508-901, Brasil

E-mail:[1] ramon@ipt.br

**Abstract.** It was developed a non-destructive testing system based on magnetic methods for characterization of steel specimens, used in calibration of Charpy impact testing machines. The magnetic properties saturation, remanence, coercivity, and the hysteresis curves were used to create a "magnetic signature" of reference to ensure the value of energy absorbed by these standard specimens.

## 1. Introduction

The Charpy impact test is widely used in mechanical construction industries for verification of mechanical properties of structural steel [1]. Through this test is determined the energy absorbed by the fracture of a sample with well-defined geometry. The amount of the absorbed energy reveals the ability of material to resist to dynamic loads, shock and defines safe operation temperatures (maximum operational temperature lower than the ductile-brittle temperature). The more ductile is the fracture, the greater the energy absorbed. Low energy indicates high hardness, but brittle materials. There is also a correlation between this energy and the tensile strength. The geometry of the specimen, as well as the machinery and procedures for conducting the tests is specified in technical standards [2]. A schematic drawing of the equipment is shown in Figure 1. These machines have to be periodically calibrated [3], and special specimens (standard specimens) are used for this purpose

Through heat treatments it is possible to change the microstructure of the material in order to modify the energy absorbed in the process of rupture. Brittle fractures are obtained in hard materials, as in martensitic steels, while ductile fractures occur in steels with predominantly ferritic structure. The certified test specimens come from metrological institutions like the NIST [4] and ERM [5]. In Brazil there are still no suppliers. Efforts involving the National Metrological Institute - Inmetro and the IPT of São Paulo are being carried out for the development of these reference materials [6]. These test specimens were manufactured for testing machines with swinging hammers schematically shown in Figure 1. The rectangular bars with 10 mm x 10 mm x 55 mm are machined with a V-notch, where it begins to break up due to stress concentration. The schematic drawing of this specimen is shown in Figure 2. Standard specimens should be characterized properly to ensure the value of the absorbed energy at break. The microstructure must also be checked in addition to the dimensional characterization. This microstructure is defined mostly by the chemical composition of the steel and by heat treatments carried out.

Journal of Physics: Conference Series **648** (2015) 012003 doi:10.1088/1742-6596/648/1/012003

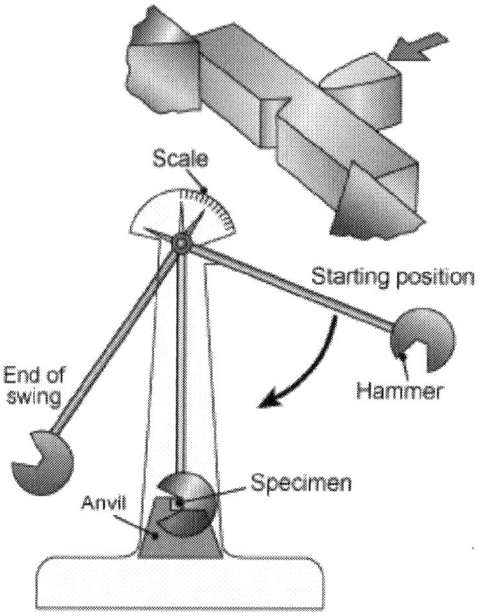


**Figure. 1.** Typical configuration of the equipment to perform Charpy testing. (Extracted from http://www.hardiepolymers.com)

The characterization of the material can be made by x-ray scattering methods, Mössbauer spectroscopy, metallographic analysis laboratory, and other techniques such as electron microscopy. In addition to the long times of measurement and sample preparation, these methods are expensive and only provide results from regions near the surface of the specimens. Alternatively, hardness tests are also used, but only at the ends of the specimens to avoid disturbing the material near the area of rupture.

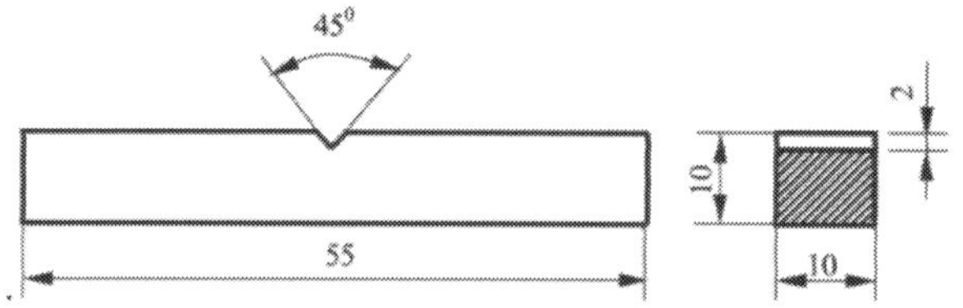


**Figure 2.** Charpy specimen (V-notch type A).

## 2. Objectives

The aim of this work is the development of non-destructive testing, based on magnetic measurements, to ensure the results of the energy absorbed in the breakage of standard specimens used in the verification of Charpy testing equipment. The implementation of this quality control is made by the rejection of specimens that feature magnetic properties off limits previously established.

## 3. Magnetic Methods

Magnetic characterization can be obtained from the information extracted from its curves of magnetization and hysteresis, relating the magnetic induction $B$ with the applied field strength $H$. Some notable points of these curves are the polarization of saturation to intense fields $J_{sat}$, the remanence $B_r$ or $J_r$ (residual magnetization when an intense applied magnetic field is zeroed), and the coercivity $H_c$ (magnetic field intensity that nulls the magnetisation). A device usually employed in this kind of characterization is the vibrating sample magnetometer (MAV). The samples used in this equipment have the shape of small needles, and the impact specimens are too big for this method. Furthermore, the MAV operates in "open circuit": the magnetic flux passing by permeability transition magnetic regions creates demagnetizing fields within the sample. That means the field intensity inside the material is smaller than the external field applied.

In this work, the magnetic characterizations were obtained by inductive technique (histeresigraph) in "closed loop", and so the field inside the sample is identical to the applied magnetic field. The specimen was positioned in contact with the polar poles of an electromagnet. The magnetic induction was measured with the aid of a coil with 50 turns wrapped around the specimen, and coupled to an electronic integrator (Laboratorio Elettrofisico – Digital Fluxmeter). The magnetic field was measured with a gaussmeter (Lake Shore 475), whose Hall probe was positioned next to the pickup coil. Figure 3 shows a detail of the specimen positioned between the polar poles of the electromagnet.

**Figure 3.** Impact specimen positioned between the polar poles of the electromagnet.

## 4. Microstructure

There is a correlation between the magnetic properties and the mechanical properties, because both are affected by the microstructure of the material. The typical microstructure of steels with low carbon content usually contains phases as pearlite and ferrite. The pearlite is a lamellar structure composed of ferrite and cementite. The final microstructures, and consequently the mechanical properties, are strongly influenced by heat treatment to which the steel are subjected. Starting from high temperatures (austenitization), the slow cooling results in ferromagnetic phases as cementite + ferrite or pearlite + ferrite. The resulting steel is more ductile, and absorbs more energy on break. In the case of the quick cooling (quenching) is also formed martensite, a ferromagnetic phase, and retained austenite (a paramagnetic phase). In this case the steel is very hard, but more brittle and less energy is absorbed at break.

## 5. Results and Discussion

The standard specimens or SPs were made of ASTM A36 steel with chemical composition C: 0.25; MN: 0, 801.20; Q: 0.04; S: 0.05; SI: 0.40. Nine SPs were tested. Three were measured in the receiving state (SPs 7, 8, 9). The other six were initially heated to 900° C. Three were kept in the oven for slow cooling in distinct rates (SPs 1, 2, 3). Two were quickly cooled down in water (SPs 5, 6), and

one was cooled down in the air (SP 4). All the SPs were broken in the Charpy impact machine after magnetic characterization tests.
It was possible to reach the magnetic saturation of the material by applying strong magnetic fields, when the magnetizing curves reach a plateau ($J_{sat}$ = 2,0 T). This limit is defined basically by the amount of alloy elements in the steel. All SPs showed saturation curves very similar to those shown in Figure 4. In these charts, the magnetic polarization $J$ is related to magnetic induction $B$ by the relation: $J = B - \mu_0 H$.

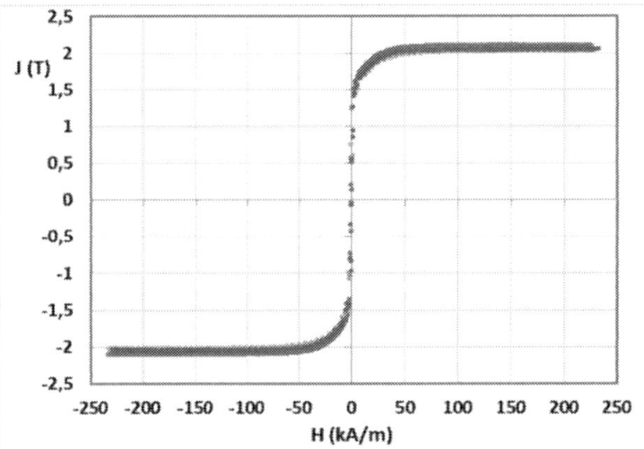


**Figure 4.** Magnetic saturation curves of the SPs under strong magnetic field intensities.

The coercivity $H_c$ and the remanence $B_r$ or $J_r$ were obtained from the hysteresis curves measured in low intensity fields range. These curves are shown in Figure 5. The SPs quickly cooled in water show less break energy and an enlargement of the hysteresis curve caused by the formation of martensite. The energy of rupture increases to the SPs slowly cooled down because there was the formation of more ductile phases, such as ferrite and pearlite.

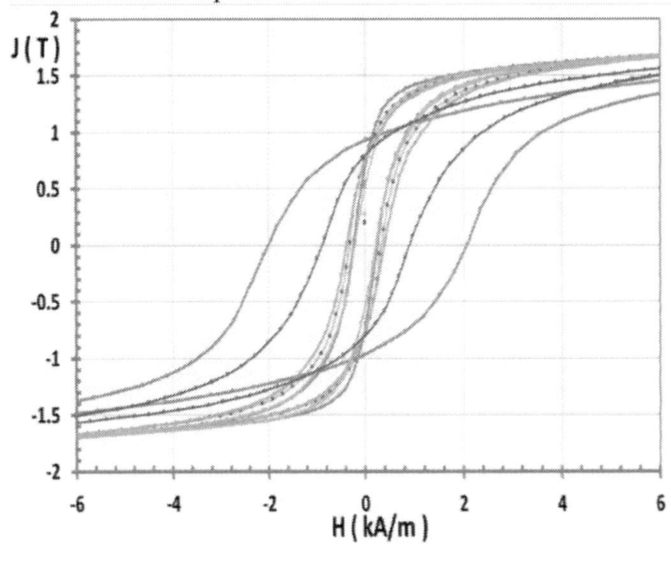


**Figure 5.** Hysteresis curves for SPs of the same chemical composition, but subject to different heat treatments. Wider hysteresis curves (red and blue) refer to the SPs quickly cooled in water.

The appearance of the rupture zone in SP 1 (cooled slowly) can be observed on Figure 6. The rupture is ductile with change of cross section. There wasn't the complete separation of the parties. Figure 7 shows the most fragile rupture that happened to the SP 6, which was quickly cooled in water.

**Figure 6.** Fracture aspect of SP with greatest rupture energy (heat-treated with very slow cooling in oven).

**Figure 7.** Fracture aspect of SP with the lowest rupture energy (heat-treated with quick cooling).

Figure 8 shows the graph of impact rupture energy as a function of remanence, while Figure 9 shows the break energy in function of coercivity for all SPs tested. The SPs cooled slowly are plotted in red, while the ones cooled quickly are in blue, and the ones without any thermal treatment are in black.

Usually, the calibration process of Charpy equipment employs SPs for three different energy bands. In this work there were arbitrarily selected three energy values: low - 50 J, average - 120 J, high - 160 J. In spite to illustrate the procedure of a possible quality control system, three tolerance bands (± 3 %) were created around this mean value of energy. These bands are indicated by the horizontal lines in figure 8 and 9.

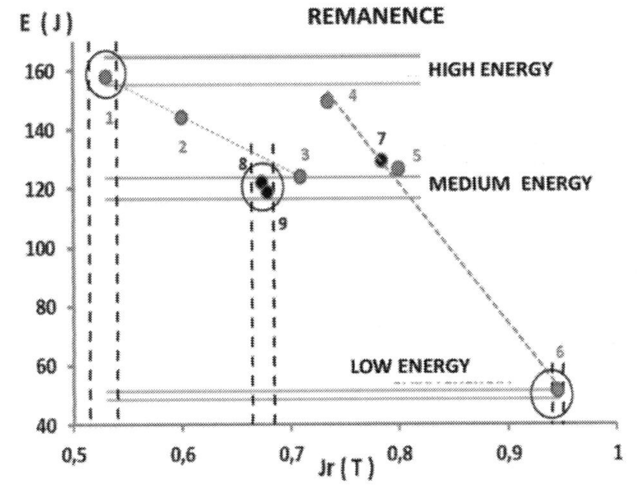


**Figure 8.** Rupture energy as a function of a magnetic property: remanence. The data were obtained from the hysteresis curves. Horizontal lines indicate the band tolerance (± 3 %) for the energies "high", "medium" and "low", while vertical lines show the tolerance ranges of remanence for acceptance of SPs on each band of energy.

Remanence and coercivity were the magnetic properties chosen as parameters to establish the criteria for acceptance of SPs due to major change values in these properties related to different heat treatments. The adopted criteria is composed by two rules: I) If the value of remanence is out of the remanence bands defined for each energy, then the SP is rejected; II) If the value of coercivity is out of the coercivity bands defined for each energy, then the SP is rejected. Graphically: Vertical dashed lines in the figures 8 and 9 show the tolerance ranges of magnetic properties for acceptance of the SPs in each energy level.

Journal of Physics: Conference Series **648** (2015) 012003 doi:10.1088/1742-6596/648/1/012003

Note that only the criterion of coercivity is not sufficient to distinguish samples of high and medium energy, as most of the tested SPc presents the coercivity between 200 A/m and 300 Am. The criterion of remanence expands this region of uncertainty, and allows to clearly exclude the SP 7 from the classification of medium energy. In terms of energy, if the tolerance band to the energy value was increased to 5 %, then SP 5 and the SP 3 could be included in the range of medium energy. This is a problem because SP 5 (martensitic), SP 3 (ferritic), and SPs 8, 9 (ferrite + pearlite) have completely different microstructures. The magnetic criteria allow differentiating the SPs by their microstructure. So, the narrower is the tolerance range, the more specific are the microstructures. This is an important issue because the microstructure can affect other properties of Charpy tests as the transition temperature between ductile and brittle rupture.

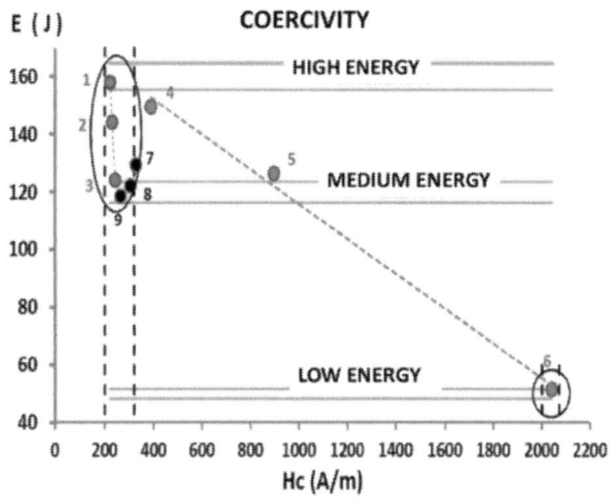


**Figure 9.** Rupture energy as a function of a magnetic property: coercivity. The data were obtained from the hysteresis curves. Horizontal lines indicate the band tolerance (± 3 %) for the energies "high", "medium" and "low", while vertical lines show the tolerance ranges of coercivity for acceptance of SPs on each band of energy.

## 6. Conclusion

Magnetic characterization tests have great sensitivity, low cost, rapid implementation, and they do not require a previous preparation of the specimens. The magnetic measurements are very affected by the microstructure of the material, and a magnetic signature obtained under the same conditions can be used as a tool for quality control to ensure the results of Charpy impact tests. Although it was not observed significant variation in magnetic saturation in SPs subjected to different thermal treatments, the magnetic properties coercivity and remanence had great variation. Values bands for remanence and coercivity were adopted to set selection criteria to ensure that the specimen is broken at the specified energy. It should be noted that the smaller is the tolerance of the magnetic values, the higher is the rate of rejection of the SPs and the greater is the specificity of their microstructures. The results indicate the feasibility of quality control based on magnetic properties, but the magnetic characterization of a far greater number of SPs should be carried out to ensure the reliability of this method.

## References

[1] ABNT NBR ISO 148-1:2013- Materiais metálicos — Ensaio de impacto por pêndulo Charpy
[2] BS EN ISO 148-3:2008 – Metallic materials – Charpy pendulum impact test.
[3] ABNT NBR ISO 148-2:2013- Materiais metálicos - Ensaio de impacto por pêndulo Charpy – Verificação de máquinas de ensaio.
[4] http://www.nist.gov/mml/acmd/structural_materials/charpy-verification-program.cfm.
[5] ERM – European Reference Materials (http://www.erm-crm.org)
[6] M. F. F. Pereira, J. A. Paz Cruz, W. Link, L.C.C. Freitas, A. C. M. Garcia, *"DEVELOPMENT AND MANUFACTURE OF CHARPY STANDARDS TEST SPECIMENS"*, Proceedings of the 18th Imeko World Congress International, Rio de Janeiro, Brazil, September 2006.

# Determination of the uncertainty of microhardness in the evaluation of hardfacing obtained by welding

**A Cruz-Crespo[1], J E S Leal[2], A Piratelli-Filho[3], T O Mendez[4] and R V Arencibia[5]**

[1,4] Central University "Marta Abreu" of Las Villas. Research Center of Welding, Road to Camajuaní, 5 ½ km Santa Clara, Villa Clara, Cuba
[2,5] Federal University of Uberlândia, 2121, João Naves de Ávila Avenue, Campus Santa Mônica, Building 5F, Zip Code: 38400-902, Uberlândia – MG. Brazil
[3] University of Brasília, Faculty of Tecnology, Department of Mechanical Engineering, University Campus Darcy Ribeiro, Asa Norte, 70910-900, Brasília, DF, Brazil

E-mail: arvaldes@mecanica.ufu.br

**Abstract**. The present work evaluates the microhardness measurement uncertainty using a Vickers indenter in the characterization of hard deposits, obtained with experimental electrode for hardfacing of cutting knives in sugar mills. The following stages were proposed: i) obtainment of the deposit with experimental electrode, ii) obtainment of the samples, iii) chemical characterization, iv) metallographic analysis, v) microhardness measurement and vi) uncertainty measurement assessment. It was observed that the obtained deposit presents chrome carbides and eutectic matrix composed of austenite and carbides. The expanded uncertainty associated to the microhardness measurement is highly influenced by the low repeatability of the obtained microhardness values.

## 1. Introduction

The sugar industry is characterized by the presence of various mechanical components subject to abrasive wear and continuous rubbing occur during manufacturing and components recovery. Among these components, the knife is an essential element that affects the milling process efficiency ratios.

Literature [1-2] reported that abrasion, corrosion and impact are wear mechanisms to which are submitted the knives' working edges, with the last one the less likely to occur nowadays due to the care and cleaning treatment of sugar cane prior to reach the feeding table, being the abrasion considered as the main wearing mechanism.

It is common practice the covering of the knives working edges using the Shielded Metal Arc Welding process (SMAW), with abrasion-resistant alloys of the type AWS 5:13 EF and Cr -A1A, which pour a (not diluted) chemical composition in weight (%) varying from *3.5%* to *4.5%* carbon, *4.0%* to *6.0%* manganese, *0.5%* to *2.0%* silicon and *20%* to *25%* chromium [3-4].

The AWS A5.13 [3] Standard establishes 26 types of surface coating consumables cast iron base (Fe) for SMAW. From these, *22* offer good resistance properties to abrasive wear, seven of which yield an alloying system with proper resistance at low pressure working conditions. The main alloying elements of such consumables are C, Mn, Si, Cr and Mo, with elements Ni (4%) and Ti (1.8%) being introduced separately [3].

The coating of knives and other components subjected to abrasion is carried upon the composition, microstructure and properties of the deposits, and also on technological criteria for applying the coating process. Note that, studies on the use of processing criteria that lead to a reliable interpretation of the results and their metrological traceability are scarce in the literature.

This work aims at determining the aspects that affect the microhardness measurement uncertainty in the characterization of hard coatings obtained with an experimental electrode applied to knives used in the sugar industry.

## 2. Methodology

The characterization of the deposit with experimental electrode applied to knives used in sugar mills [5] was carried out at the Central University of Welding Research Center Martha Abreu de Las Villas, Cuba. In order to obtain samples for chemical and structural analysis, it was used the standard SFA 5:13 [3] which establishes the size of the deposit (50 mm, 13 mm and 16 mm for electrodes with 4 mm metal core diameter). During the deposition, it was used a MANSFELD welding source with 110 A direct current and inverted polarity (DC+). The deposition was performed in multiple runs by manual welding at an average speed of 15 cm/min at room temperature.

### *2.1. Chemical analysis*

The chemical characterization was made by atomic emission spectroscopy of the sample probes that were extracted through slicing the deposit at the cross section of the top surface.

### *2.2. Metallographic analysis*

For metallographic characterization, the CIS-MET-030 [6] and CIS-MET-027 [7] preparation and samples observation procedures were followed. The samples were taken from the deposit mid portion, cutting them perpendicularly to the welding path. After that, the samples were trimmed, polished, etched with Murakami reagent and finally observed with a NEOPHOT 32 optical metallographic microscope.

### *2.3. Microhardness measurement*

The Vickers microhardness was determined using a Shimatsu microdurometer, as specified by the ASTM E92 [8] Standard. The load applied was *50 gf* during *15 s*. The imprints were made so that the spacing between them was greater than *0.5 mm*, making a total of 15 imprints. The Vickers microhardness was calculated from the diagonal length values ($d_1$ and $d_2$) formed by opposite vertices of the pyramid base left by the penetrator. The length of the diagonal was measured by the microscope`s micrometer with *400x* magnification and resolution *0.25 μm*. The expanded uncertainty associated with the calibration is *0.02 μm* for a coverage factor (*k*) equal to *2.00* and *95.45%* coverage probability. Each diagonal was measured three times.

The hardness measurement process can be considered a non-destructive test, since the tested part can be reused. However, the damage caused by the indenter in the test point makes impossible to evaluate the process uncertainty by making repeated measurements at the same location [9].

The arithmetic average of the diagonal length values was calculated by equation (1).

$$d = \frac{d_1 + d_2}{2} \tag{1}$$

For each indentation, the Vickers hardness (*HV*) was determined by equation (2), where *F* is the load applied during the test and *A* is the surface area of printing.

$$HV = \frac{F}{A} = \frac{2 \cdot F \cdot \sin\frac{136^\circ}{2}}{d^2} \approx 1.8544 \cdot \frac{F}{d^2} \tag{2}$$

### *2.4. Uncertainty measurement assessment*

The evaluation of measurement uncertainty was performed by applying the methodology proposed by ISO [10]. The main sources of uncertainty in the hardness measurement are related to the sample, hardness testing machine, the environment and the operator [9]. Some works were found on evaluation of microhardness measurement uncertainty. Among them, Ellis [11] considered the following sources of influence related to the testing machine: the linear repeatability, the load, the resolution and the indenter angle.

In this work, it was considered the following influencing factors: the linear repeatability and resolution of the machine, the load, the expansion of the measuring system and the variability of the obtained microhardness, expressed through the sample standard deviation.

Initially, the standard uncertainty associated with the length measurement of the imprint diagonals ($d_1$ and $d_2$) were calculated starting from equation (3). In this case, three influence variables were considered to determine the diagonal lenghts, $d_1$ and $d_2$.

$$d_1 = d_2 = dm + \Delta R + \Delta IC + \Delta A \tag{3}$$

In this expression: $dm$ represents the measured length of the diagonals $d_1$ or $d_2$; $\Delta R$ is the correction associated to the microscope resolution; $\Delta IC$ is the correction associated to the microscope calibration; $\Delta A$ is the correction associated to the microscope expansion.

The uncertainty associated with the load applied during the test was declared in the calibration certificate as expanded uncertainty of *0.02 gf* with *95.45%* coverage probability and coverage factor ($k$) of 2. The standardization of the expanded uncertainty was performed by means of equation (4).

$$u(F) = \frac{U(F)}{k_F} \tag{4}$$

In this expression: $u(F)$ is the standard uncertainty associated with the applied test load; $U(F)$ is the expanded uncertainty associated with the applied test load and $k_F$ is the coverage factor associated with the expanded uncertainty of the load calibration.

Next, it was determined the standard uncertainty associated with the arithmetical average of the diagonal length values ($d$), using the mathematical model of equation (1). Finally, the uncertainty associated with the Vickers hardness was determined by means of equation (2).

After applying the uncertainties propagation law in equations (3), (1) and (2), there were obtained the equations (5), (6) and (7), respectively, allowing the estimation of the combined standard uncertainty associated with the measurement of the diagonals` length values, the arithmetic mean of the diagonals` length values and the Vickers hardness.

$$u_c^2(d_1) = \left(\frac{\partial d_1}{\partial dm}\right)^2 \cdot u^2(dm) + \left(\frac{\partial d_1}{\partial \Delta R}\right)^2 \cdot u^2(\Delta R) + \left(\frac{\partial d_1}{\partial \Delta IC}\right)^2 \cdot u^2(\Delta IC) + \left(\frac{\partial d_1}{\partial \Delta A}\right)^2 \cdot u^2(\Delta A) \tag{5}$$

$$u_c^2(d) = \left(\frac{\partial d}{\partial d_1}\right)^2 \cdot u^2(d_1) + \left(\frac{\partial d}{\partial d_2}\right)^2 \cdot u^2(d_2) \tag{6}$$

$$u_c^2(HV) = \left(\frac{\partial HV}{\partial F}\right)^2 \cdot u^2(F) + \left(\frac{\partial HV}{\partial d}\right)^2 \cdot u^2(d) \tag{7}$$

However, several microhardness $HV$ measurements are usually made whose individual uncertainty is determined by means of equation (7), but it was necessary to add a correction to equation (2) as observed in equation (8), where $\Delta s(HV)$ is the correction associated with the microhardness variability. The combined standard uncertainty of these microhardness measurements is finally determined by $u_c(HV)$ in equation (9).

$$HV = \Delta s(HV) + 1.8544 \cdot \frac{F}{d^2} \tag{8}$$

Journal of Physics: Conference Series **648** (2015) 012004 doi:10.1088/1742-6596/648/1/012004

$$u_c^2(HV) = \left(\frac{\partial HV}{\partial \Delta s(HV)}\right)^2 \cdot u^2(\Delta s(HV)) + \left(\frac{\partial HV}{\partial F}\right)^2 \cdot u^2(F) + \left(\frac{\partial HV}{\partial d}\right)^2 \cdot u^2(d) \quad (9)$$

Calculating the partial derivatives in equations (5), (6) and (9), it was obtained the equations (10), (11) and (12), respectively.

$$u_c^2(d_1) = u^2(\Delta R) + u^2(\Delta IC) + u^2(\Delta A) \quad (10)$$

$$u_c^2(d) = \left(\frac{d_2}{2}\right)^2 \cdot u^2(d_1) + \left(\frac{d_1}{2}\right)^2 \cdot u^2(d_2) \quad (11)$$

$$u_c^2(HV) = u^2(\Delta s(HV)) + \left(\frac{1{,}8544}{d^2}\right)^2 \cdot u^2(F) + \left(\frac{-2 \cdot 1{,}8544 \cdot F}{d^3}\right)^2 \cdot u^2(d) \quad (12)$$

### *2.5. Standard uncertainty assessment*

The standard uncertainty linked to the microscope resolution was determined by equation (13). Therefore, a Type B assessment was performed, considering a rectangular distribution with infinite degrees of freedom.

$$u(\Delta R) = \frac{Resolution}{\sqrt{3}} \quad (13)$$

The standard uncertainty linked to the microscope calibration was determined as a Type B evaluation, considering a normal distribution and coverage factor ($k=2$) and the expanded uncertainty stated in the calibration certificate, according to equation (14).

$$u(\Delta IC) = \frac{U(\Delta IC)}{k} \quad (14)$$

The standard uncertainty linked to microscope amplification was calculated by means of a Type B evaluation, considering a rectangular distribution and infinite degrees of freedom. In this case, a distortion was adopted due to the lens magnification corresponding to $0.01\%$ of $d_1$ and $d_2$ length values, respectively, as shown in equation (15).

$$u(\Delta A) = \frac{0.01\% \cdot d_1}{\sqrt{3}} \quad (15)$$

The standard uncertainty associated with the variability of the Vickers microhardness ($\Delta s(HV)$) was calculated by means of Type A evaluation, considering a t-Student distribution, and n-1 degrees of freedom, where n represents the number of performed indentations or the number of microhardness measurements, equation (16).

$$u(\Delta s(L)) = \frac{s(L)}{\sqrt{n}} \quad (16)$$

## 3. Results and discussion

The chemical composition of the deposit is shown in table 1. It was observed high content of carbon and chromium, usually characterized by high presence of carbides, being this material suitable for applications in the presence of abrasive wear.

**Table 1.** Chemical composition of the deposit, in weight %.

| Electrode | C | Cr | Mn | Si |
|---|---|---|---|---|
| Experimental | 3.27 | 15.34 | 4.09 | 2.78 |

Many authors, even when in unbalanced conditions as in the actual solidification process, use the ternary system Fe-Cr-C aiming at generally predicting the phases present in white casting alloys rich in chromium, commonly used in abrasive wear. Among them, Chia-Ming et al. [12] address a study of the microstructure and abrasive wear in coating of hypereutectic alloys with different carbon contents, obtained with self-protecting electrode. Also Hsuan-Han et al. [13] discuss a study of the microstructure of the multilayer coatings of hypoetectic alloys, obtained via GTAW with transversal oscillation.

Figure 1 shows the liquid projection of the rich in iron vertex of the Fe-Cr-C system, where the point A is situated, corresponding to the deposit of table 1, considering only the elements of the ternary system. It was observed that the point is located in the border of the austenite and $M_7C_3$ chromium type carbide phases, which also contains iron in the composition ($(Cr, Fe)_7C_3$). The deposit observed microstructure is shown in figure 2.

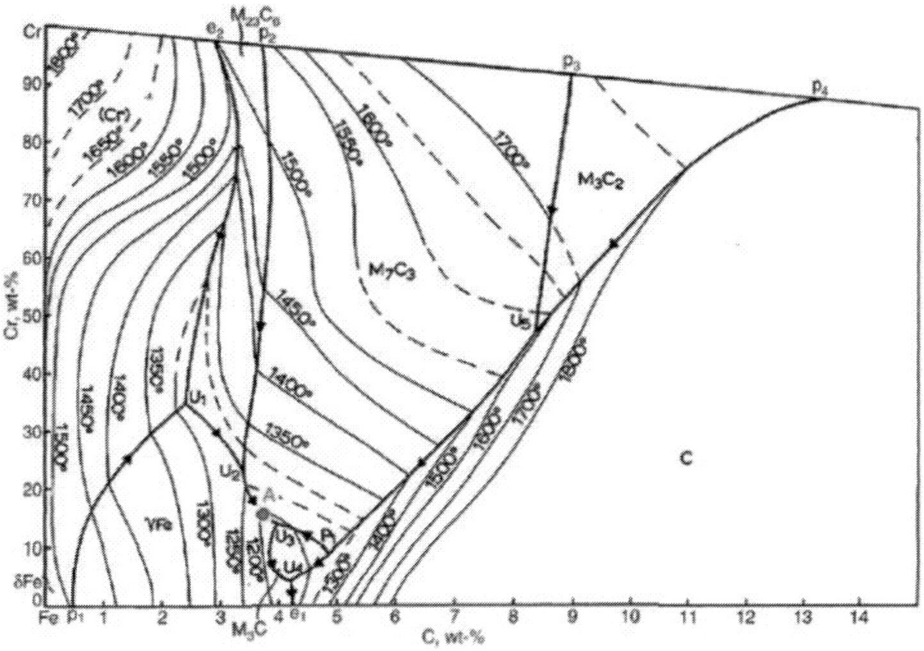


**Figure 1.** Liquids projection of ternary system Fe-Cr-C to the region rich in iron [12] and [13].

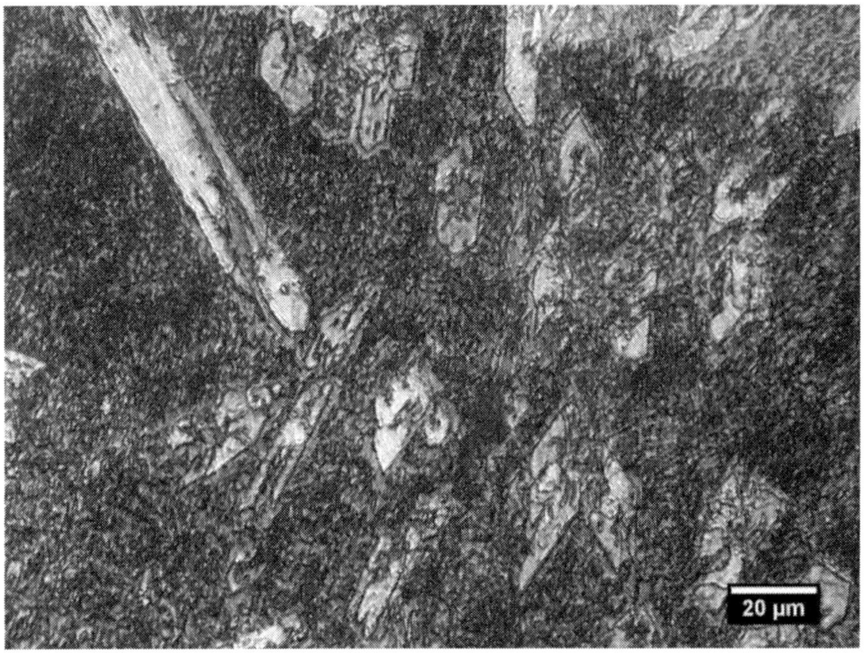


**Figure 2.** Deposit microstructure.

Figure 2 shows the presence of hexagonal carbides of $M_7C_3$ type, a noticeably eutectic region (consisting of austenite phases and $M_7C_3$ carbides) and a clear zone, ascribed to an eutectic region with greater presence of the austenite phase. Thus, the observed microstructure is fully consistent with the analysis of the phases present at the base of the balance ternary system of figure 1.

The presence of primary carbides within an eutectic matrix, which is composed of austenite and small carbides, ensures good resistance to predominating abrasive wear in knives used in sugar mills. In this sense, the result is similar to that reported in the literature for sugar industry applications. For a coating of high-chrome hypereutectic alloy obtained by SMAW, the presence of $M_7C_3$ type chromium carbides in a eutectic matrix presents characteristics suitable for abrasive wear [14-15].

In figure 3, the average hardness values, with a standard deviation of 68.27% probability, obtained in each zone of the deposit and the corresponding sample standard deviation are shown.

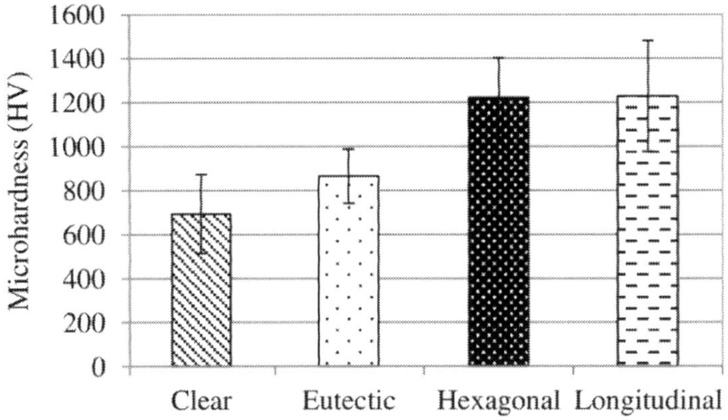


**Figure 3.** Microhardness mean values of the deposit zones (HV).

Note that in figure 3 the longitudinal cross-section of the deposit's carbides presents the highest microhardness value of *1228 HV*, which is relatively similar to the hexagonal section with. On the other hand, the eutectic region exhibits a hardness of *864 HV* while the clear zone presents *693 HV*.

The average microhardness values coincide with the values commonly reported by several authors, including Chia-Ming et al. [12], who in a study of several coatings with different carbon contents, obtained by welding with tubular electrode, obtained carbides and eutectic microhardness values in a range within which are the values achieved in this study.

It is worthy of note that the identified clear (white) regions also have eutectic matrix, but as they solidify too late get lesser amount of carbides and higher amount of austenite, probably this fact was also influenced by the pro-austenitic effect of manganese present in the deposit (table 1).

With respect to measurement repeatability, it was observed that, for a *68.27%* probability, the longitudinal area has a standard deviation of *252 HV*, while the eutectic and hexagonal areas exhibit lower values as *179 HV*. In the literature are generally referred high microhardness dispersion values of microstructural constituents of white high chrome castings.

For example, Stevenson and Hutchings [16] in a study of the wear behavior of various coatings, show high variability of carbides microhardness values, with standard deviation of *210 HV*, while the eutectic matrix displays *125 HV* standard deviation.

The footprint of the imprint left by the Vickers indenter is small, allowing the hardness determination of individual micro constituents of a microstructure, however the small size of the imprint contributes to a lower repeatability of the measurement.

The repeatability of the hardness measurement can be significantly influenced by the operator, depending on his training, thoroughness and skill [17]. The operator is the one who determines the boundaries of the diagonal imprints during measurement with microscope. Thus, according to the method used, the parameter "operator skill" has a significant influence on the final result of the measure, and it may represent a major source of error and consequently of uncertainty [18].

Next, the Chauvenet criterion was applied to eliminate extreme values (outliers). In this case were identified only 2 microhardness values that do not follow the mainstream and so were discarded, which are highlighted in table 2 in gray. Again, the arithmetic mean and the standard deviation were calculated.

By eliminating the extreme values, the repeatability of measuring hardness was improved in the clear and the hexagonal zones, showing a reduction of about *48%* and *16%*, respectively.

It was also observed in table 2 that the microhardness values corresponding to the deposit's clear zone have two perfectly distinguishable populations of values, with much lower values than those expected in the austenite phase; therefore, it can be deduced that there are minute dispersed carbides imperceptible under optical microscopy. In turn, the eutectic zone shows appropriate values for this constituent, in correspondence with the values of the alloying elements.

On the other hand, the microhardness of the carbides shows appropriate values, but it should be noted that their high dispersion may be a possible consequence of heterogeneous crystallization caused by the particular conditions of cooling, characteristic of the welding deposition process and the alloy composition that lead to an extreme imbalance state. The heterogeneity of the carbides microhardness values was associated to the heterogeneity of their own chemical composition as a result of heterogeneous nucleation.

**Table 2.** Microhardness values of the deposit (HV) after applying Chauvenet criterion.

| | Clear | Eutectic | Hexagonal | Longitudinal |
|---|---|---|---|---|
| 1 | 633 | 726 | 1095 | 897 |
| 2 | 689 | 766 | 1049 | 909 |
| 3 | 700 | 1006 | 946 | 927 |
| 4 | 810 | 810 | 1254 | 965 |
| 5 | 714 | 857 | 1197 | 1027 |
| 6 | 557 | 841 | 1283 | 1095 |
| 7 | 1254 | 726 | 1072 | 1145 |
| 8 | 739 | 909 | 1283 | 1254 |
| 9 | 633 | 946 | 1524 | 1283 |
| 10 | 603 | 985 | 1254 | 1346 |
| 11 | 473 | 1095 | 1225 | 1396 |
| 12 | 753 | 655 | 1605 | 1486 |
| 13 | 689 | 927 | 1314 | 1524 |
| 14 | 540 | 946 | 1225 | 1564 |
| 15 | 603 | 766 | 1005 | 1605 |
| Average | 653 | 864 | 1195 | 1228 |
| Standard deviation | 92 | 123 | 150 | 252 |

Figure 4 shows the imprint area values left by the Vickers indenter in the first measurement area in each zone of the evaluated deposit. As expected, a clear zone has the highest area value (*0.023* $mm^2$), while the Longitudinal zone has the lowest value (*0.012* $mm^2$).

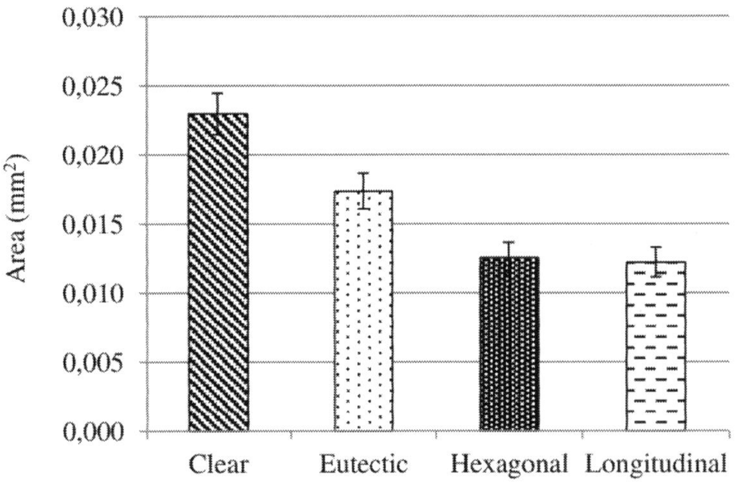


**Figure 4.** Area values and the associated combined standard uncertainty (68.27%).

With respect to the combined standard uncertainty associated with the area measurement, it was observed that it takes similar values for the four evaluated zones, with *0.0015* $mm^2$ for the Clear zone, *0.0013* $mm^2$ for the Eutectic and *0.0011* $mm^2$ for the Hexagonal and Longitudinal zones.

Figure 5 shows the expanded uncertainty values associated with the microhardness measurement in the four considered zones for *95%* coverage probability.

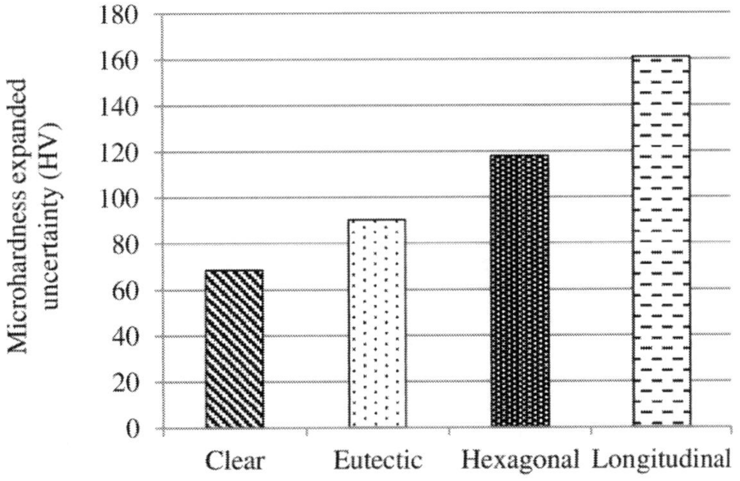


**Figure 5.** Expanded uncertainty values (95%) associated with the microhardness measurement.

In figure 5, it can be observed that the expanded uncertainty associated with the microhardness assume values of *68 HV*, *90 HV*, *118 HV* and *161 HV* for the Clear, Eutectic, Hexagonal and Longitudinal zones, respectively. These figures follow the trend observed for the standard deviation.

The variable that contributed most to the final uncertainty measurement in all cases was the variability of the obtained microhardness values. Note that for the clear zone, the contribution of this variable was *52%*, while for the eutectic zone was *50%*. In turn, the readings variability was responsible for *46%* and *68%* of the final uncertainties of the hexagonal and longitudinal zones.

## 4. Conclusions

The uncertainty of microhardness measurements is highly influenced by the standard deviation of microhardness values, even after applying Chauvenet criterion for extreme values withdrawal.

The presence of $M_7C_3$ type carbides in a eutectic matrix (carbides + austenite) are indicators of good resistance to abrasive wear of the knives used in sugar mills. The presence of austenite becomes softer the matrix, preventing the release of the carbides.

**Ackonowlegments**
The authors are grateful to FAPEMIG, CNPq and CAPES for financial support.

## References

[1] Dulón J 2003 *Manual de Soldadura para la Industria Azucarera* ed. Talleres de Impresiones gráficas del Ministerio del Azúcar (La Habana) p 116.
[2] Hugot E 1980 *Manual para Ingenieros Azucareros* Editorial Pueblo y educación (La Habana) p 802.
[3] SFA 5.13, 2000 Parte C, Section II (AWS A5.13-2000) *Specification for Surfacing Electrodes for Shielded Metal Arc Welding*, pp 305-126.
[4] Weman K 2003 *Welding process handbook* (New York: CRC Press LLC, ISBN 0-8493-1773-8)
[5] Ortiz T M et al. 2007 *Electrodo Tubular Revestido para la Reparación de Piezas Sometidas a Desgaste Abrasivo en el Sector Azucarero* (Centro Azúcar, n. 2).
[6] Procedimiento CIS-MET-030 2002 *Preparación de muestras* Centro de Investigaciones de Soldadura CIS-UCLV.
[7] Procedimiento CIS-MET-027 2002 *Observación de muestras por microscopia óptica* Centro de Investigaciones de Soldadura CIS-UCLV.

[8] ASTM E92 2010 Standard Test Method for Vickers Hardness of Metallic Materials.
[9] EURAMET 2011 Guidelines on the Estimation of Uncertainty in Hardness Measurements EURAMET cg-16 Version 2.0 p 22.
[10] ISO TAG 4/WG 3 2008 *Guide to the Expression of Uncertainty in Measurement* (Geneva Switzerland) p 141.
[11] Ellis R A 2003 *Uncertainties of Vickers Hardness Test Blocks* (xvii IMEKO World Congress) June 22 – 27, Dubrovnik, Croatia, pp 982-985.
[12] Chia-Ming Ch, *et al.* 2010 *Microstructure and wear characteristics of hypereutectic* Fe–Cr–C *cladding with various carbon contents* Surface & Coatings Technology **205** (doi:10.1016/j.surfcoat.2010.06.021) pp 245–250
[13] Hsuan-Han L *et al.* 2014 *Effect of Oscillating Traverse Welding on Microstructure Evolution and Characteristic of Hypoeutectic Hardfacing Alloy* Surface & Coatings Technology **239** (http://dx.doi.org/10.1016/j.surfcoat.2013.11.048) pp 233–239.
[14] Buchanan V E, Shipway P H and McCartney D G 2007 *Microstructure and Abrasive Wear Behaviour of Shielded Metal Arc Welding Hardfacings Used in The Sugarcane Industry Wear* **263** (doi:10.1016/j.Wear.12.053) pp 99–110.
[15] Buchanan V E, McCartney D G and Shipway P H 2008 *A Comparison of The Abrasive Wear Behaviour of Iron-Chromium Based Hardfaced Coatings Deposited by SMAW and Electric Arc Spraying Wear* **264** (doi:10.1016/j.wear.2007.04.008) pp 542–549.
[16] Stevenson A N J and Hutchings I M 1995 *Wear of Hardfacing White Cast Irons by Solid Particle Erosion Wear* pp186-187, pp 150-158.
[17] Podchibiakin D et al. 2003 *Comparación de Diferentes Métodos de Medición de la Impronta de la Imagen de Microdurómetros* (Jornadas Sam/Conamet/Simposio Materia) pp 1979-82.
[18] Gabauer W 2000 S M & T Standards Measurement & Testing Project No. SMT4-CT97-2165 *Manual of Codes of Practice for the Determination of Uncertainties in Mechanical Tests on Metallic Materials*, Code of Practice No. 14, The Estimation of Uncertainties in Hardness Measurements, p 16.

# Impact of Pitot tube calibration on the uncertainty of water flow rate measurement

**Icaro de Oliveira Buscarini, Andre Costa Barsaglini, Paulo Jose Saiz Jabardo, Nilson Massami Taira, Gilder Nader**

Instituto de Pesquisas Tecnologicas do Estado de Sao Paulo - IPT, Av. Prof. Almeida Prado, 532, São Paulo, SP, 05508-901, Brazil

E-mail: buscarini@ipt.br, barsaglini@ipt.br, pjabardo@ipt.br, nmtaira@ipt.br, gnader@ipt.br

**Abstract.** Water utility companies often use Cole type Pitot tubes to map velocity profiles and thus measure flow rate. Frequent monitoring and measurement of flow rate is an important step in identifying leaks and other types of losses. In Brazil losses as high as 42 % are common and in some places even higher values are found. When using Cole type Pitot tubes to measure the flow rate, the uncertainty of the calibration coefficient ($C_d$) is a major component of the overall flow rate measurement uncertainty. A common practice is to employ the usual value $C_d = 0.869$, in use since Cole proposed his Pitot tube in 1896. Analysis of 414 calibrations of Cole type Pitot tubes show that $C_d$ varies considerably and values as high 0.020 for the expanded uncertainty are common. Combined with other uncertainty sources, the overall velocity measurement uncertainty is 0.02, increasing flowrate measurement uncertainty by 1.5 % which, for the Sao Paulo metropolitan area (Brazil) corresponds to $3.5 \times 10^7$ m$^3$/year.

## 1. Introduction

Loss of treated water is large in Brazil. While there are no official measurements, the losses are estimated to vary between 37 and 42 %. These losses are usually a result of leaks and illegal water connections. To improve this picture, it is necessary to implement an efficient measurement/monitoring in all stages of the supply network.

Since flow rate is an important parameter in quantifying the losses, it is important to use calibrated flow meters with low uncertainty. Usually, large flow meters are calibrated in field by mapping the velocity profile [1] using Cole type Pitot tubes (Figure 1). Thus, flow rate uncertainty is directly dependent upon the uncertainty of the velocity measurement. High flow rate uncertainty not only, potentially, corresponds to the large monetary value of unmeasured water flow but also makes it difficult to identify losses in the water supply system, whether they are physical in nature such as leaks or non-physical such as illegal connections to the supply network, for instance.

The advantage of using a Cole type Pitot tube when mapping the velocity profile is that it can be used in water pipes with with a wide range of sizes, from diameters as low as 300 mm up to a few meters. While the mean value of the calibration coefficient or discharge coefficient ($C_d$) of a Cole type Pitot tube is 0.869 [2], its use has a large impact on flow rate uncertainty as shown here, where 414 calibrations were carried out between 2004 and 2013 at the anemometry lab of the Instituto de Pesquisas Tecnologicas (IPT), Brazil.

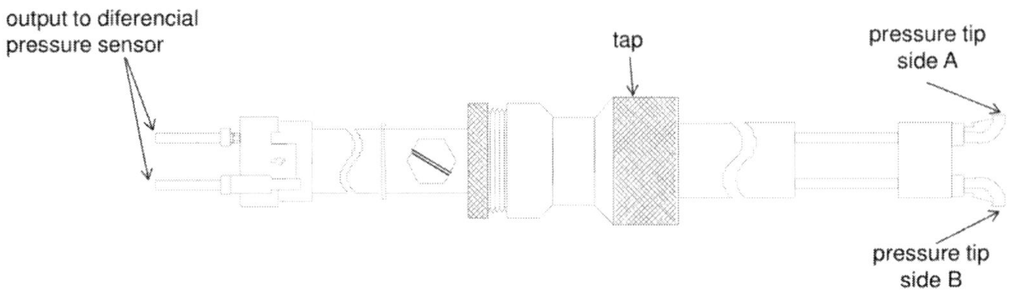


**Figure 1.** Schematics of a Cole type Pitot tube.

The impact of the uncertainty of $C_d$ was estimated for the metropolitan area of São Paulo (RMSP). RMSP encompasses 8 distribution systems of treated water: Alto Cotia (1.2 $m^3/s$), Alto Tietê ($15\,\mathrm{m^3/s}$), Cantareira ($33\,\mathrm{m^3/s}$), Guarapiranga ($14\,\mathrm{m^3/s}$), Rio Claro ($4\,\mathrm{m^3/s}$), Rio Grande ($5\,\mathrm{m^3/s}$), Baixo Cotia ($0.9\,\mathrm{m^3/s}$) and Ribeirão da Estiva ($0.1\,\mathrm{m^3/s}$). This amounts to $73.2\,\mathrm{m^3/s}$ or $2.3 \times 10^9\,\mathrm{m^3/year}$ of clean water. It should be noted that even with small changes, on the order of 1 %, of the calibration coefficient, can lead to errors of several million cubic meters at the end of one year.

Analyzing this database of calibrations shows that the periodic calibration of Cole type Pitot tubes is important. This is not a simple metrological quality system requirement [3] but a fundamental step if improvement of flow rate uncertainty is desired.

## 2. Flow rate measurement using Cole type Pitot tube

The Cole type Pitot tube is a differential pressure device designed by Edward Cole in 1896 [2], that has a total pressure tap (tip A, faces upstream) and a wake pressure tap (tip B, faces downstream) as shown in Figure 1.

When Cole type Pitot tubes are used for measuring water flow rate, the velocity profile is measured using parameters established in the Standard ISO 3966 [1]. Figure 2 shows an example of velocity profile including the measurement points.

The flow rate ($Q$) is determined using the mean velocity ($V$) obtained from the velocity profile along one or two orthogonal diameters in a cross section of the pipe with area $A$:

$$Q = V \cdot A \tag{1}$$

The mean velocity ($V$) is obtained from point measurements of velocity ($V_i$) at positions specified by the Standard ISO 3966 [1]. The velocity ($V_i$) in each point is obtained from the differential pressure across the Cole type Pitot tube according to the following equation:

$$V_i = C_d \sqrt{\frac{2\Delta P}{\rho}} \tag{2}$$

where $C_d$ is the calibration coefficient of the Cole type Pitot tube, $\Delta P_i$ is the pressure differential measured in each point $i$ of the profile and $\rho = 998.202\,\mathrm{kg/m^3}$ is the density of water at $20\,°\mathrm{C}$ and $101.325\,\mathrm{kPa}$.

The mean calibration coefficient recommended by the literature for conventional Cole type Pitot tubes (a slight modification of the usual Cole type Pitot tube will also be analyzed), including corrections is 0.869 [2]. Figure 3 presents a set of measurements that are commonly

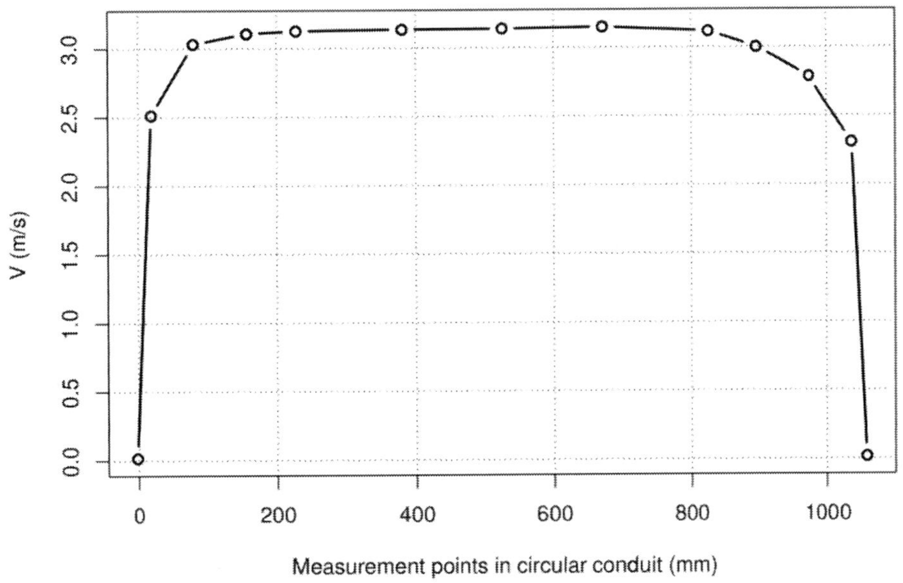


**Figure 2.** Velocity profile.

used by water utility companies in Brazil [4]. This figure shows the dependence of $C_d$ with Reynolds number. The Reynolds number is defined as:

$$Re = \frac{VL}{\nu} \tag{3}$$

where $L$ is a characteristic length, here fixed as 1 $m$, $\nu = 1.004 \times 10^{-6}\,\mathrm{m/s^2}$ is the kinematic viscosity of water at $20°C$.

The results in figure 3 for $0.5 \times 10^6 \leq Re \leq 3.0 \times 10^6$, correspond to a velocity range of $0.5 \leq V \leq 3.0\ m/s$ in water. In this velocity range, the calibration coefficient of the Cole type Pitot tube varies between 0.883 for 0.5 m/s and 0.861 for 3.0 m/s with a mean value of 0.867.

The modified Cole type Pitot tubes tested in this work possesses a safety pin between the tips (figure 4) to protect the tips from damage when internal walls of the pipes are hit. Calibrations of modified Cole type Pitot tubes over the years resulted in a mean calibration coefficient of 0.886.

Table 1 presents the sources of uncertainty of the flow rate measurement using Pitot tubes. The standard uncertainties associated with these sources are estimated from instrument calibration and measurement record. These uncertainties are combined using the following equation:

$$u_c^2(y) = \sum_{i=1}^{n} [c_i]^2\, u^2(x_i) = \sum_{i=1}^{n} \left[\frac{\partial f}{\partial x_i}\right]^2 u^2(x_i) \tag{4}$$

where $x_i$ is an uncertainty source, $c_i$ is its sensitivity coefficient and $u(x_i)$ is the corresponding standard uncertainty. The uncertainty of the calibration coefficient is one of the major contributions to the overall uncertainty of the the velocity [5]. Thus, the use of a single mean value for $C_d$ can have a large effect on the uncertainty of flow rate.

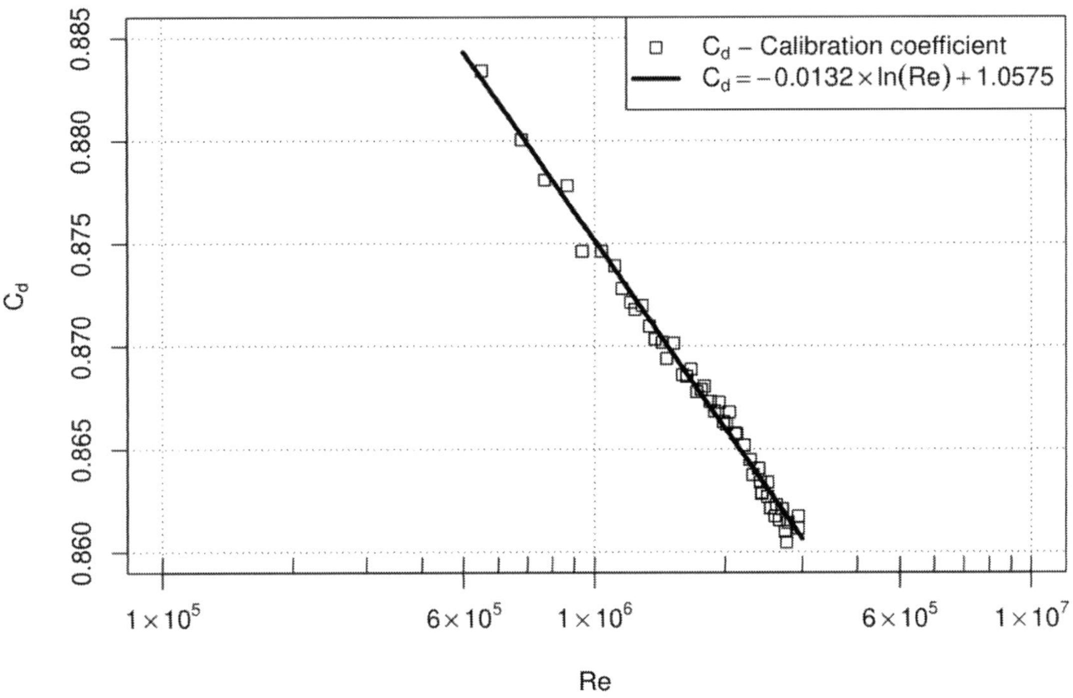


**Figure 3.** Reynolds number dependence of calibration coefficient ($C_d$) according to [4].

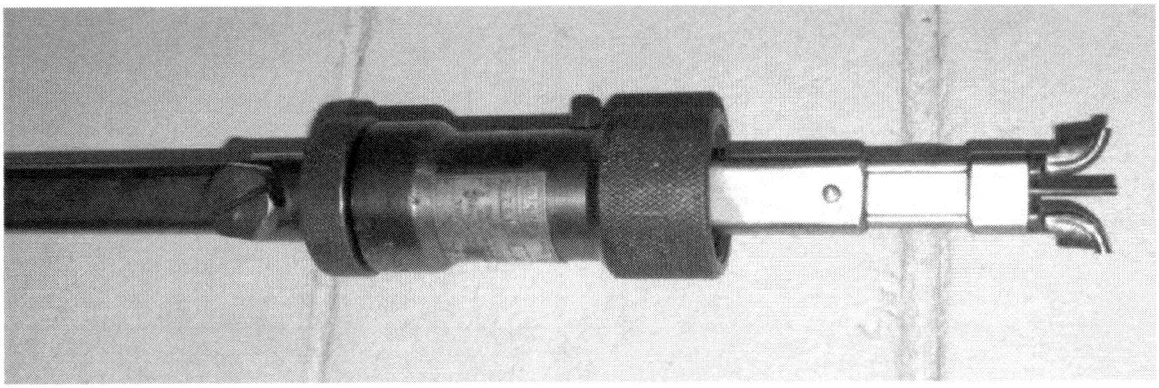

**Figure 4.** Photograph of a modified Cole type Pitot tube with pin between tips.

At IPT, the Cole type Pitot tubes are calibrated using an aerodynamic wind tunnel (figure 5) [6]. Calibrations in a wind tunnel of Cole type Pitot tubes to be used in water agree with calibrations in Pipes and towing tank (both executed at IPT) [7].

During calibration, the Cole type Pitot tubes are positioned at the discharge of the wind tunnel. A static Pitot tube is used as a standard. Both Pitot tubes are connected to manometers and two rising sequences of points consisting of 10 air velocities between 5 and 36 m/s are compared. Using Reynolds similarity, where

$$Re_{water} = Re_{air}$$

these velocities correspond to 0.3 and 2.4 m/s in water respectively.

**Table 1.** Uncertainty sources and its sub-components in flow rate measurement using Pitot tubes.

| Uncertainty Source | Components |
|---|---|
| Local velocity | Pressure differential |
| | Density |
| | Calibration coefficient |
| | Velocity time fluctuations |
| | Velocity gradient |
| | Pitot tube orientation |
| | Pitot tube blockage |
| Mean velocity | Local velocity |
| | Pitot tube position |
| | Numerical integration |
| Pipe cross section | Diameter |

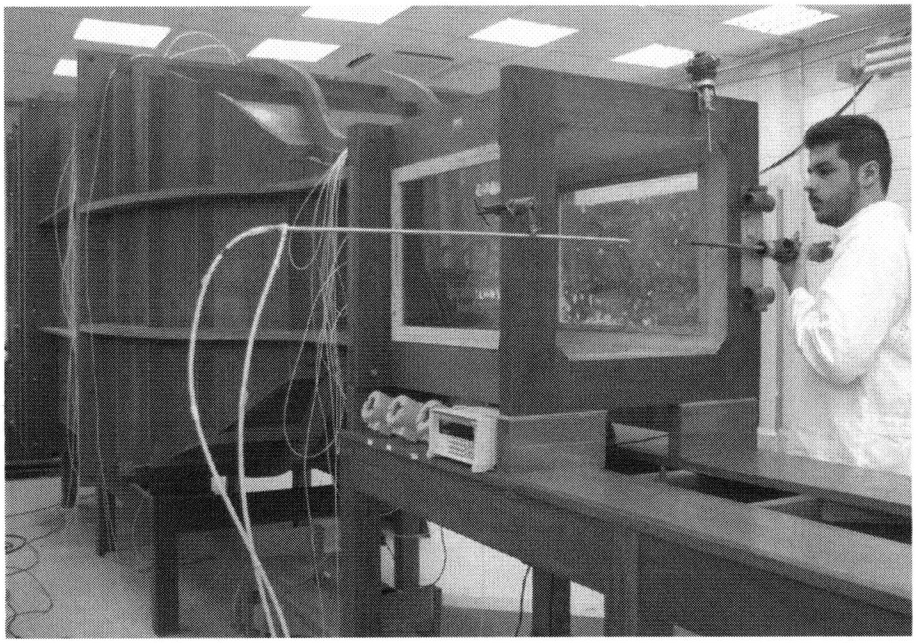

**Figure 5.** Calibration of Cole type Pitot tube at the aerodynamic wind tunnel.

The calibration coefficient of the Cole type Pitot tube is obtained using the following equation:

$$C_d = C_p\sqrt{\left(\frac{\Delta P_P}{\Delta P_C}\right)} \quad (5)$$

where $C_p = 0.997$ is the calibration coefficient of the static Pitot tube used as a standard [8], $\Delta P_P$ and $\Delta P_C$ are, respectively, the pressure differential of the static and Cole type Pitot tubes.

## 3. Methodology

The variation of calibration coefficients of Cole type Pitot tubes were analyzed from a total of 414 calibrations consisting of 212 conventional Cole type Pitot tubes and 202 modified Cole type Pitot tubes (with a protective pin between tips). These calibrations date from 2004 until 2013.

Initially, the mean and standard deviation of calibrations were obtained for both modified and conventional Cole type Pitot tubes. Then, the successive calibrations of four individual Pitot tubes were analyzed. Each Pitot tube analyzed was calibrated 5 times and the conventional Cole type Pitot tubes are identified by C1 and C2, while the modified Pitot tubes are identified by M1 and M2.

While the calibration coefficient ($C_d$) varies with Reynolds, usually most variation happens at lower Reynolds numbers (low speeds). The calibration coefficient is calculated for each calibration using the mean $C_d$ of calibration points in the region of lower dependence on Reynolds number (larger velocities). The calibration coefficients obtained from every calibration was used to estimate mean value and standard deviation of $C_d$ for both conventional and modified Cole Pitot tubes. This information was used to determine the expanded uncertainty of $C_d$ with a confidence level of approximately 95 %. This overall uncertainty was compared to the individual uncertainty of each calibration. Finally, the impact on the flow rate from using a single generic value for $C_d$ instead of a calibrated value was estimated.

## 4. Results

### *4.1. Individual calibrations*

For some Pitot tubes, several calibrations over a period of time where available. With this information it was possible to verify how $C_d$ changed over the lifetime of the Pitot tube due to typical wear and damage. This change is mostly due to deformations of the tips that usually occur when the tips hit the inside walls of pipes.

Two conventional (C1 and C2) and two modified (M1 and M2) Cole type Pitot tubes which were calibrated 5 times over the span of 10 years were analyzed. Figure 6 shows the mean $C_d$ for Pitot tubes C1 and C2 for different Reynolds number (Re). Each point corresponds to the average of 5 calibrations and the vertical bars represent the range of $C_d$ values obtained considering every calibration of the same Pitot tube at the same Re.

Ideally, the Cole type Pitot tube is symmetrical so that either tip can be used to measure the total pressure. Unfortunately manufacturing problems and wear cause differences on the tips and both sides are calibrated (A and B). In figure 6(b), the differences between sides A and B are considerable and therefore, special care should be taken to identify the sides during calibration. Figure 7 corresponds to modified Pitot tubes M1 and M2.

Figures 6 and 7 show that even for a single Cole type Pitot tube there is significant standard deviation of the order of 0.007 of $C_d$ during its life time. Variations of the same order are also observed when comparing calibrations of both sides of the Cole type Pitot tubes. In the case of Pitot tube C2 this difference is close to 0.013 while for M2 this difference is 0.007, lower but still considerable. These differences are close to the uncertainty itself: different sides of a same Pitot tube are equivalent to completely different Pitot tubes.

### *4.2. Historical database of Cole type Pitot tube calibration coefficients*

The 212 conventional Cole type Pitot tubes presented a mean value of $\overline{C}_d = 0.8696$ with a standard deviation of 0.0094 as shown in figure 8(a). The mean calibration coefficient is close to the value of 0.869 originally suggested by Cole [2]. However, this historic mean has a large uncertainty that can not be neglected.

The modified Cole type Pitot presents a slightly higher calibration coefficient of 0.883 and standard deviation of 0.0084 as shown in figure 8(b). This standard deviation is similar to the conventional Pitot tube.

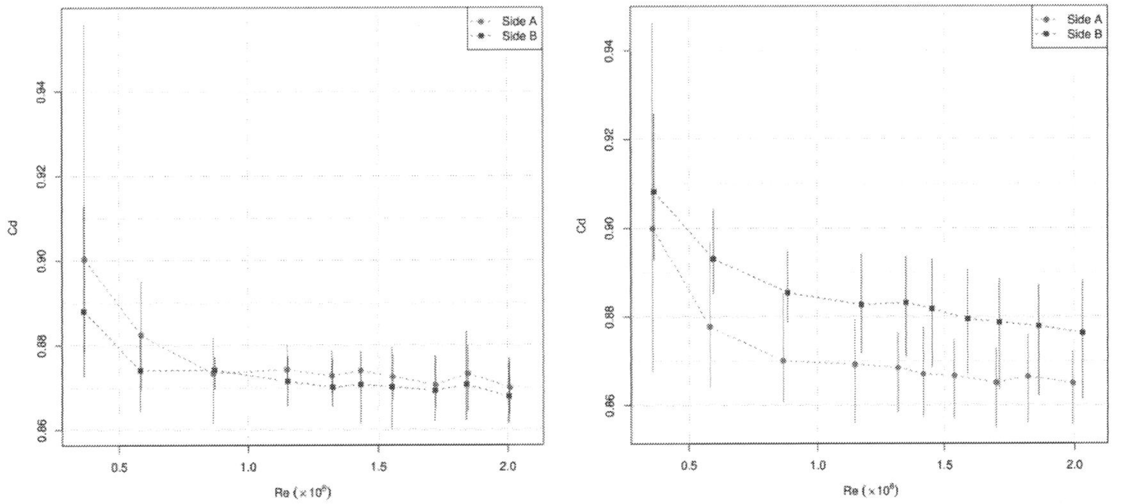


(a) C1 (b) C2

**Figure 6.** Mean value of $C_d$ and variation during 5 calibrations for sides A and B of Cole type Pitot tubes C1 and C2.

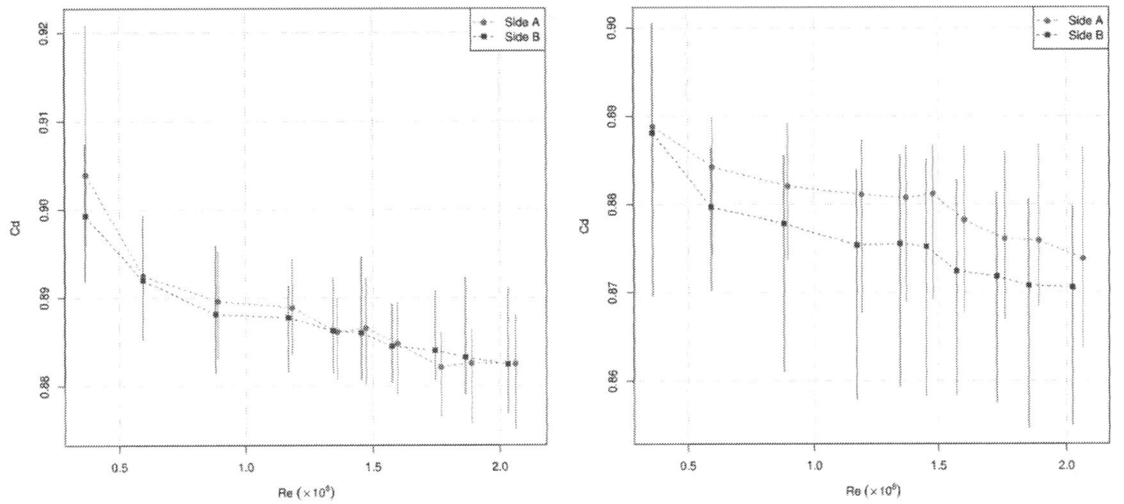


(a) M1 (b) M2

**Figure 7.** Mean value of $C_d$ and variation during 5 calibrations for sides A and B of Cole type Pitot tubes M1 and M2.

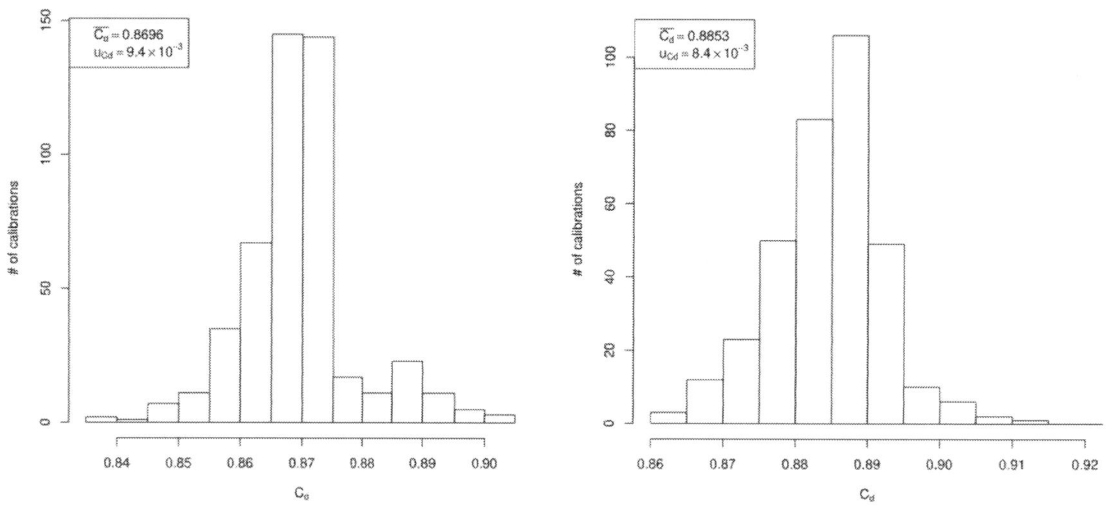


(a) Conventional Cole type Pitot (b) Modified Cole type Pitot

**Figure 8.** Histogram of the calibration coefficients in historical database.

This historical database of modified and conventional Cole type Pitot tubes include periodical calibrations of a large number of distinct Pitot tubes. Considering only Pitot tubes that were calibrated at least 3 times, and calculating the standard deviation of $C_d$ for each Cole type Pitot tube, we obtain the histogram in figure 9(a).

This histogram has a peak near 0.005 and falls off rapidly standard deviations larger than 0.01. Clearly this distribution is not normal. It is interesting to remember that one of the situations when the Pitot tube is calibrated occurs when the tip is damaged and the specific Pitot tube is effectively a different Pitot tube. This histogram includes data from Pitot tubes that were repaired and Pitot tubes that suffered normal wear. Without further information on the history of each individual Pitot tube, no further assessments of the causes of the peculiar shape of the histogram can be made.

Figure 9(b) obtained by integrating the histogram of figure 9(a) shows the probability that a given Pitot tube will have a standard deviation of $C_d$ during its lifetime lower than a certain value. For a probability of 95 %, this value corresponds to 0.105. It is worthy of note that this value is very close the standard deviation of all calibration coefficients (figure 8).

### *4.3. Impact on flow rate measurement*

When the velocity profile is well behaved as shown in figure 2 and considering all sources of uncertainty in table 1, equation (4) can be used to obtain the expanded uncertainty of flow rate.

If a calibrated $C_d$, with expanded uncertainty of 0.008, is used, the expanded uncertainty of the flow rate is $U_Q = 2.5$ %. On the other hand, if a generic mean value for $C_d$ is used with expanded uncertainty of 0.02, the overall expanded flow rate uncertainty id $U_Q = 4.0$ %.

Therefore, the use of a calibrated Pitot tube reduces the overall uncertainty of the flow rate by approximately 1.5 %. This reduction is still expected, approximately, for poor velocity profiles that present a higher overall uncertainty.

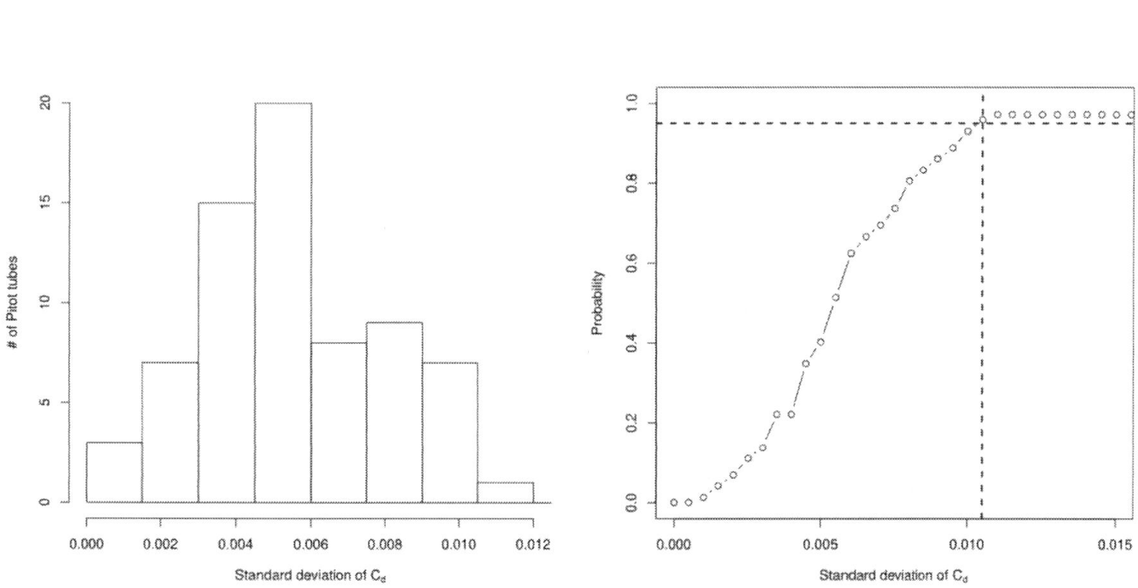


(a) (b)

**Figure 9.** (a) Histogram of standard deviation over the life of single Cole type Pitot tube calibration coefficients of tubes calibrated at least 3 times; and (b) probability that a given Pitot tube will have, over several calibrations, a standard deviation lower than a given value, calculated by integrating the histogram in part (a).

The contribution of calibration coefficient uncertainty is important but still limits the uncertainty of flow rate measurement to values above 2 %. However by using a non calibrated Cole type Pitot tube an additional 1.5 % is to be expected. While some may consider this value, in general, acceptable, it is large enough to affect water loss management and the financial impact on supplier or consumer can not be dismissed. As an example, the city of São Paulo distributes $73.2\,m^3/s$ of treated water. An additional uncertainty of 1.5 % represents $3.5 \times 10^7\,m^3/year$.

## 5. Conclusions

While the use of mean calibration coefficient for Cole Pitot tubes seems reasonable since this mean doesn't change over a period of decades, when comparing individual Cole type Pitot tubes, the historical database with 414 calibrations carried out from 2004 until 2013 shows that the expanded combined uncertainty of the mean calibration coefficient can be over 0.02.

This uncertainty contributes to an additional 1.5 % uncertainty which corresponds to $3.5 \times 10^7\,m^3/year$ in the metropolitan region of São Paulo. Periodic calibration of the Pitot tube reduces this uncertainty to 0.008 and is an essential step if lower flow rate uncertainties are desired.

## References

[1] ISO 3966 2008 Measurement of fluid flow in closed conduits - velocity area method using pitot static tubes
[2] Cole E and Cole E S 1939 *Trans. ASME* 465–75
[3] ISO/IEC 17025 2005 General requirements for the competence of testing and calibration laboratories
[4] CETESB Sao Paulo *Manual de pitometria*
[5] Kang W 2013 Experimental and numerical investigations of factors effect on the s-type pitot tube coefficients *Proc. 16th Int. Flow Measurement Conf.* (FLOMEKO)
[6] Nader G, Santos C, Jabardo P J S, Cardoso M, Taira N M and Pereira M T 2006 Characterization of low turbulence wind tunnel *IMEKO XVIII World Congress and IV Brazilian Congress of Metrology*
[7] Espirito Santo G, Taira N M and Santos C 2004 Comparison of calibration of pitot tubes in a wind-tunnel, towing tank and pipe. private Communication at IPT of results not published
[8] ISO/IEC 61400 2005 Wind turbine generator systems – part 12: Wind turbine power performance testing annex f

# Automated gauge block pair length difference calibration and associated uncertainty sources

**W Oliveira Jr[1] and R S França[1]**

[1]Laboratory of Interferometry (Laint), Optical Metrology Division (Diopt), Directory of Scientific and Industrial Metrology (Dimci), National Institute of Metrology, Quality and Technology (Inmetro), Av. Nossa Sra. das Graças 50, Duque de Caxias, RJ, Brazil, Cep 25250-020
E-mail: wojunior@inmetro.gov.br

**Abstract**. A reduction for interferometric uncertainties in length difference at gauge block pairs is presented. An automated processing designed to compensate geometric fringe visualization effects and four-alternate wringing technique are used to achieve small combined uncertainties for length difference calibrations, maintaining a good compliance with the EAL-G21 determinations.

## 1. Introduction
According the EURAMET cg-2 standard [1] the dimensional metrological community must have at their disposal several calibrated standards in form of a gauge block pairs set, to be used as a necessary electromechanical comparator measurement validation tool. The reduced uncertainties considered in this reference are essential prerequisites for these standards and they can be achieved by high-accuracy interferometric methods. Use of automated phase estimation techniques is specially indicated to reduce the measurement uncertainties to the recommended levels.

## 2. Objective
One of approaches to maintain the quality assessment and extend the traceability chain of the calibration systems is to compare results of same standards obtained with older and updated systems [2]. To avoid operator biases and coarse direct visual observation uncertainties obtained as in the original interferometric system, it was developed an automated image capture computer system. This system was originally based on specific software designed in Visual Basic to process phase information of interferograms that were obtained by an attached high sensitivity astronomical digital camera, with near square aspect ratio (with resolution corresponding to 512 x 512 pixels) and good time exposure control. This last feature allows it to work together with a small luminous flux $^{114}Cd$ spectral lamp to produce their wavelength references as length standards.

### 2.1. *Interferometric length difference measurements*
In differential measurements, as the gauges must be commonly affected by changes of optical paths at the block sides in absolute interferometric techniques, taking a simple numeric subtraction between two absolute lengths of a gauge pair measured in separate moments carry all combined uncertainties due to thermal expansions and air refractive index effects. To avoid that increase of uncertainty values we must measure directly the length difference of both upper surfaces of gauges in a pair by

interferometric methods. This approach allows us to reduce all length-dependant components to sub-nanometer amounts in pairs of similar nominal lengths.

We must compare the phases in fringe patterns of both upper surfaces to estimate their length difference values. But the intrinsically asymmetric nature of geometric point-of-view in comparing lengths from one upper surface center point to another precludes us of using a mean value of two reference points to compensate for slight fringe inclinations as made in absolute symmetrical approaches. In 2000 [3] there was designed a compensation method, consisting in alternating choose of both reference surfaces ("straightening" the fringes over these surfaces) and taking the average of those two measurements, ignoring orientation at base fringes. That can produce a large dispersion sometimes. An estimation error can be directly perceived by inspection at fringe fractions in the figure below:

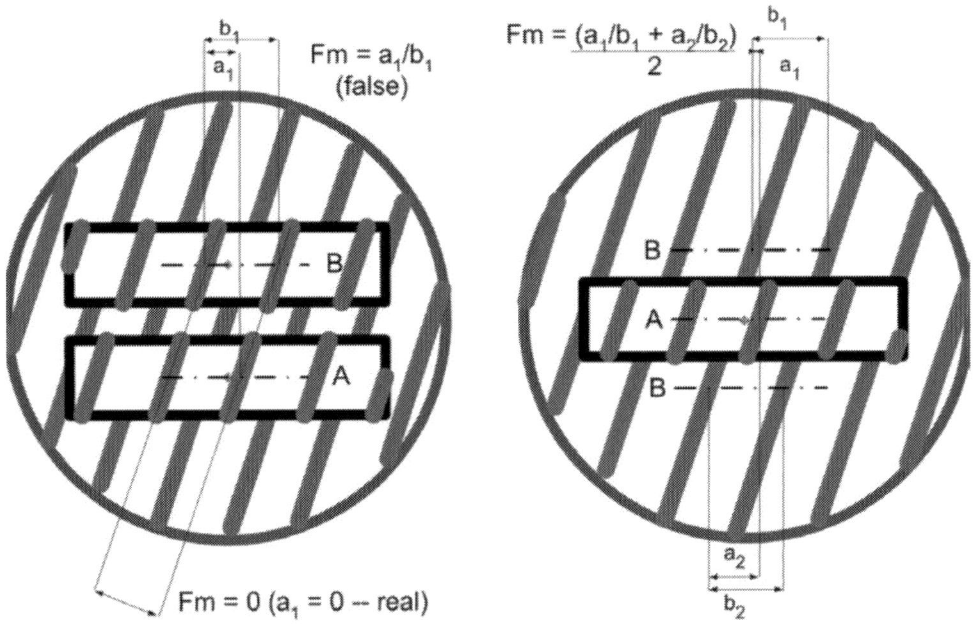


**Figure 1.** Fraction fringe error due to misalignment of fringe pattern inclination with connecting line of central gauge block points (left image: differential method; right image: absolute method).

## 3. Measurement results

The first report presented in [3] was obtained from good surface quality gauge block pairs, measured through visual estimation. Those results were good enough to validate a new calibration service. The original software measurement techniques were initially aimed to be used in absolute interferometric gauge block calibrations, but there were afterwards directly adapted to perform differential measurements.

A posterior study for estimation of uncertainty component due to dispersion obtained after measurements of some wrings at both blocks of a pair was made by changing position/wringing of only one block at a time, alternating and inverting sides (but not its wringing surface) to produce four independent geometric configurations in relative positions over same wringing surfaces as shown in figure 2. The base fringes were maintained strictly vertical to avoid huge misaligning effects. That should barely compensate for minimal asymmetric flatness deviations and effects of positional wringing. The original proposal was take two measurements each wringing, inverting the reference surface to reduce the misalignments effects, as shown before in figure 1.

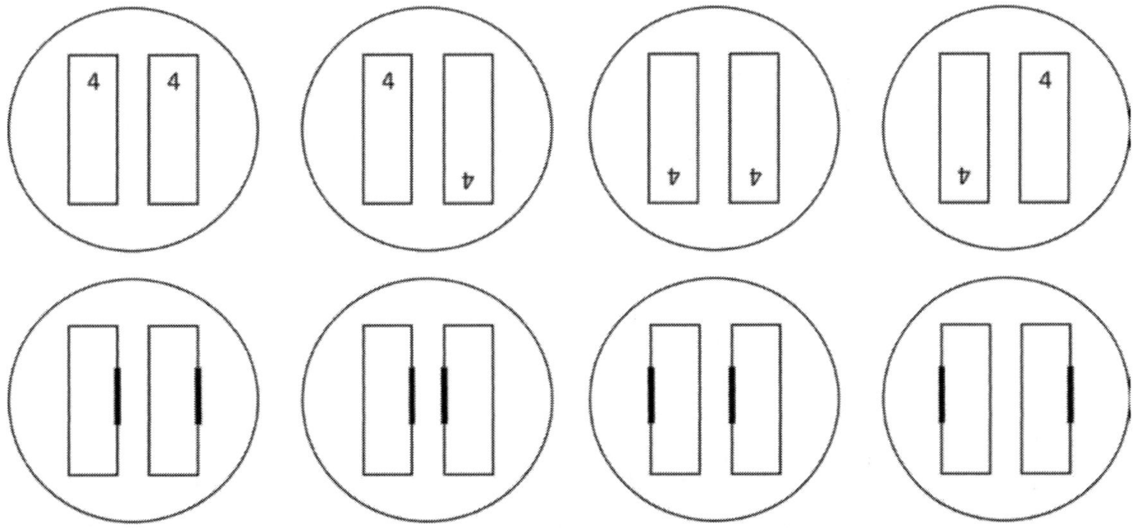


**Figure 2.** Upper view of four positional wrings for a complete measurement of the length difference of a gauge block pair; Row above: 4 mm gauge block pair; Row below: 100 mm gauge block pair (similarly for blocks larger than 6 mm) – Lateral body inscriptions depicted as thick traces.

### 3.1. *Phase estimation software for difference measurements*

In order to reduce the uncertainty component due to visual estimation of the fringe fractions an automated method was developed based on interferometric image processing obtained from a CCD camera, as used in absolute gauge-block length calibrations. This method just allowed user-controlled channel definitions, because both parallel 1D interferogram channels must cross the central point of upper surfaces in both gauges to perform length difference calibrations. The phase estimation was originally based on simple maxima enhancement skeletonizing methods and direct pixel counts between maxima to estimate the phase difference between both channels.

The figure 3 below shows the main window of the software developed in Visual Basic for automated phase difference estimations in four different wavelengths, obtained from the $^{114}$Cd spectral lamp as length reference. The information of fraction fringes in more than one wavelength is essential to determine length deviations in nominal lengths (or difference in lengths) that are much larger than half wavelength as defined in [4] and a numerical coincidence of four lengths must be achieved, combined as in equation (1) below.

$$L = (N_i + f_i)\frac{\lambda_i}{2} \tag{1}$$

Where $N_i$ is the integer number of half wavelengths ($\lambda_i/2$) contained in the length $L$ and the $f_i$ is the fractional fringe read, for each sequenced wavelength selected.

As two of the gauge pairs defined by [1] have length differences of circa 10 μm and 5 μm, this sequence combination of fringe readings should be performed to achieve the unequivocal values of length deviation measurements.

Journal of Physics: Conference Series **648** (2015) 012006 doi:10.1088/1742-6596/648/1/012006

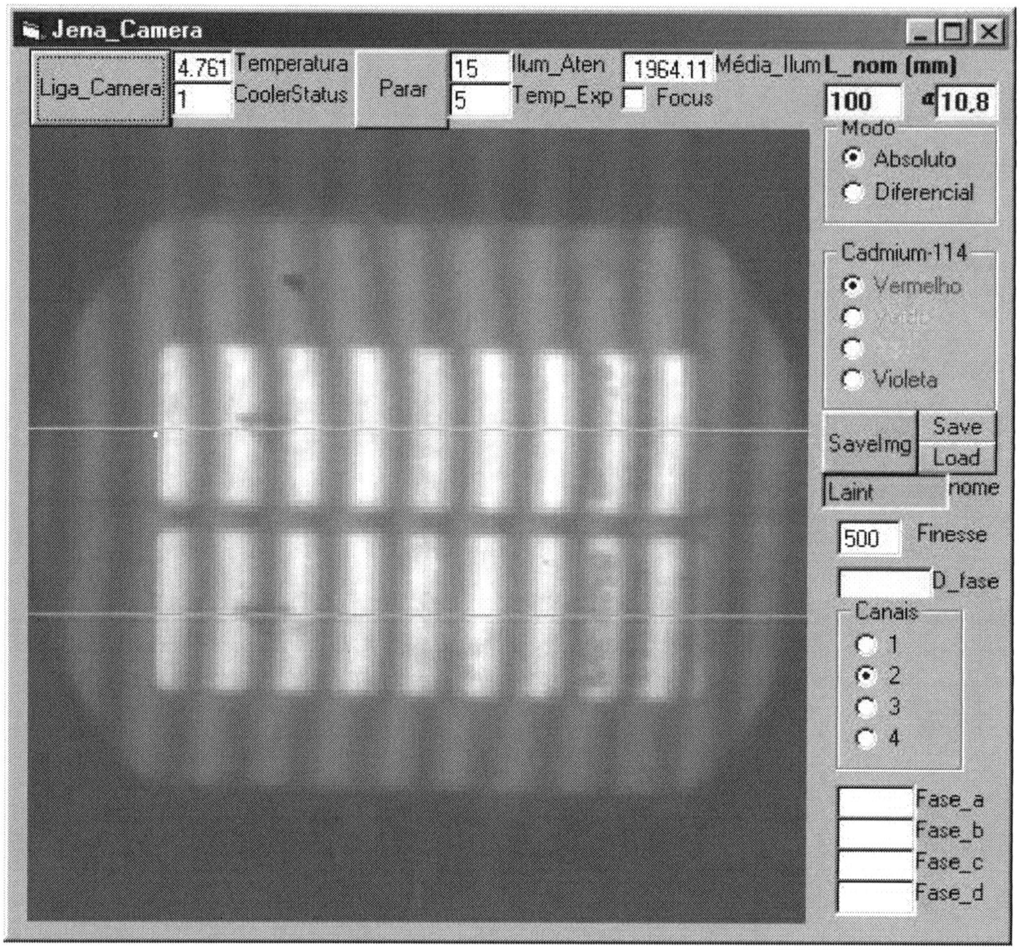


**Figure 3.** Measurement and inspection main window of software designed to estimate phase differences in interferograms. The image inside shows a monochromatic interferogram taken from a 4 mm steel gauge block pair wrung on a quartz plate, seen from above.

After the first automated method for phase estimation was implanted, the older calibration procedure was refurbished, replacing the "composite" result of two measurements, i.e. resulting from the operation ((A-B) – (B-A))/2 described in [3], by taking one valid measurement only. A special cautious must be taken in maintaining the base fringes as orthogonal as possible to avoid huge misaligning effects. Below, in table 1, there are depicted a set of main uncertainty components used in a typical uncertainty budget of a pair difference length calibration by interferometric method. The fraction fringe estimation uncertainty component presented here is derived from the pixel count method and it has a value more than 2 times smaller than uncertainties due to the older visual estimations.

**Table 1.** Updated Uncertainty Budget for length difference of gauge blocks with similar nominal length (in this case: 100 mm).

| Uncertainty Components | *u(xi)/nm* | $c_i$ | *u(l)/nm* |
|---|---|---|---|
| Fraction fringe estimation | 0.02 | 132.5 | 2.60 |
| Wringing (2x) | 2.90 | 1.414 | 4.10 |
| Flatness/Parallelism (2x) | 2.80 | 1.414 | 4.00 |
| Software resolution | 0.58 | 1.000 | 0.58 |
| Dispersion (best 4 meas.) | 2.50 | 1.000 | 2.50 |
| Wavefront errors | 3.00 | 1.000 | 3.00 |
| **Total Expanded Uncertainty: U=15 nm (k=2)** | | | |

The "best-case" expanded uncertainty, as shown in last row of table 1, is in fair agreement with declared CMC PTB pairs calibration uncertainty at BIPM database (10-14 nm, k = 2). To achieve such reduced values some homogeneity assumptions were taken in account, as the similarity at standards material features of both gauge-blocks of one pair. The uncertainties components due to slight differences of expansion coefficients and surface finishing were deemed as negligible. Also the complete temperature stabilization of standards and measurement system was assumed, to avoid the occurrence of thermal gradients inside and between the blocks or along the optical paths.

### 3.2. *Review of Uncertainty Components*

Some doubled components presented in table 1, as those due to wringing and flatness/parallelism, are composed of similar effects in both blocks of the pair. The first one took in account both wringing films between gauge blocks and wringing platen and the second one is due to respective maxima deviations of geometrical tolerances. As a common source we can inspect the software length resolution. This component is due to the steps taken by a fitting error function at the length deviation calculus, that is designed to compare ideal model results to the actual four independent fraction fringe readings (from different wavelengths). The wavefront error is another common factor caused by optical effects at optical paths and elements inside the interferometric measurement apparatus. The quoted dispersion "type A" or reproducibility component is obtained after the four differential measurements taken in different positional wrings, as presented in figure 2.

Further reductions in some of its greater components are yet to be expected. For example, that due to "flatness/parallelism" component, originally estimated in abstract dependence of maxima gauge block class tolerances and centering definitions as defined in [5], can be computed with the actual features of each individual pair. A wavefront error compensation for each interferometer type can be also implemented in automated form, measuring good reference standard surfaces as software image compensators before measuring the gauge pair. A finer step search size can be chosen to produce a direct reduction in the software length resolution component. Better fringe estimation algorithms, as suggested before, can still reduce its respective uncertainty component fivefold by using commercial phase step systems [6] or more than tenfold by using sophisticated Fourier or recent stochastic methods. In [7] it was described a possible further upgrade of the phase estimation software by adopting evolutionary computation methods and preprocessing techniques of noise reduction. This upgrade can reduce further the global uncertainty values for many types of interferometric calibrations.

Some the before deemed "negligible" uncertainties components were recently updated from results of tests made on the same measurement system: The asymmetric uncertainty contribution of second wringing of the pair (3.1 nm) was revealed to be only slightly different of the before adopted, obtained from the homogeneity assumption (2.9 nm for both wrings). The new composition of effects from first and second wringing at uncertainty budget was similar as the previous value used. A study over the uncertainty component due to the phase differences between surfaces of blocks purchased from same manufacture and made from same material (steel) and delivered within short time intervals was estimated to be around 0.98 nm. The component due to difference length variations within the assumed temperature measuring range of the interferometry laboratory could not be experimentally discriminated from other uncertainty effects and we can, therefore, still consider all temperature inhomogeneities as negligible.

## 4. Discussions and conclusions

Actual interferometric differential measurements systems must take into account some geometric and material constraints in order to attain expanded uncertainties of circa 15 nm for each gauge block pair length calibration.

A visual old-fashioned observation and estimation of fringe fractions was clearly inappropriate for accurate measurement of fringe fractions commonly found in differential measurements. Therefore, an automated image capture system was developed in Inmetro. That system combined a new developed approach, using a dispersion component produced by four-wringing positional compensation technique, with the compensated geometric misalignments and algorithmic automated estimations of fringe fractions. Those practices allowed the Interferometry Lab in Inmetro to calibrate high-accuracy dimensional standards as prime quality grade K or 00 gauge-block pairs towards uncertainty levels there are in better conformity to EURAMET cg-2 [1], and with lower uncertainty values than could be achieved by composing their individual gauge length measurements. This technique allows us to reach fairly similar values to the uncertainty values declared in PTB, the german National Metrology Institution where originally the idea for this kind of interferometric measurements was developed, aiming to establish a fair maintenance of the dimensional traceability chain, and in order to follow good metrological standards.

As future developments for this interferometric system were designed an accurate and low-uncertainty evolutionary algorithm for phase estimation as described in [7], and an upgrade of digital camera system together with its operating software, added to an advanced optical capture system. With these upgrades, the calibration services could achieve good compliance with the best metrological practices [1] and good traceability towards more technological and advanced applications [8].

## 5. References

[1] EURAMET e.V. Technical Comitee for Length 2011 *Calibration of Gauge Block Comparators - Calibration Guide, EURAMET cg-2, version 2.0 (March, 2011)*

[2] França R S and Souza V H C 2007 Length Measurement Comparisons by Interferometric Method in Brazil *Proc. COBEM 2007, 19th International Congress on Mechanical Engineering (Brasilia, Brazil, November 5-9, 2007)*

[3] Belaidi H, França R S and Massone C A 2000 Calibration of standard gauge blocks by differential interferometric method. Application to electromechanical comparators *Proc. 1st International Conference on Metrology – Trends and Applications in Calibration and Testing Laboratories (Jerusalem, Israel, May 16-18, 2000)* pp 348-354

[4] Doiron T and Beers J 1995 *The Gauge Block Handbook* (NIST Monograph 180)

[5] Decker J E and Pekelsky J R 1997 Uncertainty evaluation for the measurement of gauge blocks by optical interferometry *Metrologia* **34** 479-493

[6] Narumi T, Shutoh S, Nakamura T, Iida J and Sawabe M 1998 Development of New Automatic Gauge Block Interferometer up to 250 mm *SPIE Conf. on Recent Developments in Optical Gauge Block Metrology (San Diego, CA)* **3477**

[7] França R S and Mesquita A C 2011 Use of Applied Evolutionary Computing Techniques in Interferometric Fringe Fraction Estimation *Proc. Metrologia 2011 – 6º Congresso Brasileiro de Metrologia (Natal, Brazil, September 27-30 2011)*

[8] NASA 2010 *NASA Measurement Quality Assurance Handbook – Annex 3 Measurement Uncertainty Analysis Principles and Methods (Approved 2010-07-13, 2010)* ed. NASA-HDBK-8739.19-3

Journal of Physics: Conference Series **648** (2015) 012007 doi:10.1088/1742-6596/648/1/012007

# Uncertainty of angular displacement measurement with a MEMS gyroscope integrated in a smartphone

**Maurício de Campos Porath, Ricardo Dolci**
Universidade Federal de Santa Catarina - Campus Joinville, Rua Dr. João Colin 2700, 89218-035 Joinville/SC, Brazil

E-mail: mauricio.porath@ufsc.br

**Abstract.** Low-cost inertial sensors have recently gained popularity and are now widely used in electronic devices such as smartphones and tablets. In this paper we present the results of a set of experiments aiming to assess the angular displacement measurement errors of a gyroscope integrated in a smartphone of a recent model. The goal is to verify whether these sensors could substitute dedicated electronic inclinometers for the measurement of angular displacement. We estimated a maximum error of 0.3° (sum of expanded uncertainty and maximum absolute bias) for the roll and pitch axes, for a measurement time without referencing up to 1 h.

## 1. Introduction

Inertial navigation systems composed of tridimensional acceleration sensors (accelerometers) and tridimensional rotation rate sensors (gyroscopes) are typically used for the assessment of position and orientation in scenarios with poor GPS signal availability [1,2]. More recently, low cost inertial sensors based on MEMS (microelectromechanical systems) technology are also widely applied in consumer electronics such as smartphones, tablets and video games. An application for this type of sensor is the determination of the orientation of a smartphone, and thus switch between landscape and portrait display.

Low-cost MEMS inertial sensors are currently a popular research topic. Several researchers around the world are deriving new applications for this kind of sensors, developing solutions to improve the characterization (calibration) process and investigating the performance of the sensors [3,4,5,6,7,8,9]. However, there is a lack of proposals for the estimation of the final measurement uncertainty of these sensors. Furthermore, no performance investigation of sensors integrated in a smartphone could be identified. In this paper we will address the performance investigation of a gyroscope sensor integrated in a smartphone from a measurement uncertainty approach, with focus on angular displacement measurements.

Gyroscopes provide an output signal proportional to the rotation rate of the object. The angular displacement of the object can be obtained by integrating this signal [1,2]. Based on the results of initial experiments, we formulated the hypothesis that the accuracy of the orientation data provided by a gyroscope integrated in a smartphone of a recent model may be sufficient for applications where one would typically use a dedicated electronic inclinometer. A typical application is the determination of the inclination of a surface in relation to a given reference system. As target we defined a maximum error of 0.1°. This is a typical value for electronic inclinometers available in the market such as the digital protractor Mitutoyo 950-315 [10].

Journal of Physics: Conference Series **648** (2015) 012007 doi:10.1088/1742-6596/648/1/012007

Some advantages of the application of gyroscope sensors integrated in a smartphone are the low cost and the possibility of wireless data transfer by means of WiFi, GPRS or Bluetooth.

In this paper we present the results of a set of experiments conducted to evaluate the overall metrological performance and estimate the maximum error of a MEMS gyroscope, integrated in a high-end smartphone, for the measurement of angular displacement in the pitch and roll axes (figure 1) for low-dynamic applications. We did not evaluate the yaw axis due to the lack of an appropriate reference system.

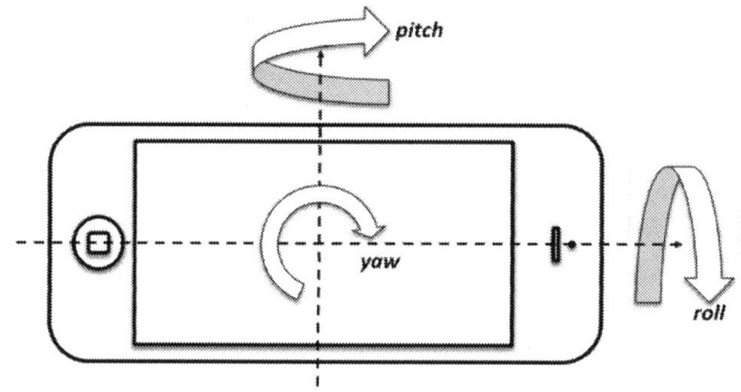


**Figure 1.** Definition of the gyroscope angles

The maximum error was estimated for the measurement range between -20° and 20° in each axis. Maximum error $E_{max}$ is defined by equation (1), where $U_{max}$ is the maximum expanded measurement uncertainty for a coverage probability of 95.45% and $|b|_{max}$ is the maximum absolute bias. The measurement uncertainty was estimated according to the "Guide to the Expression of Uncertainty in Measurement - GUM" [11].

$$E_{\max} = U_{\max} + |b|_{\max} \tag{1}$$

## 2. Equipment and software

In our experiments we used an Apple iPhone 5 smartphone (model MD297DN/A) running iOS 7.0.4. The measurement data was transferred to a laptop via WiFi network using the OSC (Open Sound Control) protocol. On the smartphone side we used the GyrOSC application, which is commercially available [12]. This application allows the transmission of the raw data of several sensors of the smartphone (accelerometer, gyroscope, magnetometer etc.) with a sampling rate of 40 Hz.

The reception and storage of the data on the laptop was realized with an application developed by us in Processing programming language [13].

For the determination of the bias curve and the repeatability of the sensors we used a calibrated sinus table and a set of calibrated gauge blocks. The sinus table presents a distance between cylinder axis of 250 mm and a leveling height of 12.5 mm. Gauge-blocks of nominal length 12.5 mm, 15 mm, 20 mm, 30 mm, 40 mm, 50 mm, 75 mm and 100 mm were used, resulting in inclinations in respect to the reference surface of the sinus table of 0.00°, 0.57°, 1.72°, 2.87°, 4.01°, 6.32°, 8.63°, 14.48° and 20.49°.

## 3. Mathematical model of the measurement process

The first step for the estimation of the uncertainty of a measurement result is to define a mathematical model that describes the measurement process. We assumed a simple linear model as described by equation (2). In this equation $\theta_{ind}$ is the displacement indicated by the sensor, $\theta$ is the actual angular displacement and $\delta_{rep}$, $\delta_{drift}$, and $\delta_{temp}$, are errors caused by the natural repeatability of the measurement process, by the drift of the signal related to the integration of the rotation rate signal and errors caused by thermal effects, respectively. The term $\delta_{ref}$ represents errors related to the uncertainty of the reference measurement system used to assess the bias (sinus table and gauge blocks). The bias is represented by the term $b$.

$$\theta_{ind} = \theta + \delta_{rep} + \delta_{drift} + \delta_{temp} + \delta_{ref} + b \tag{2}$$

## 4. Experimental quantification of the measurement errors

In this section we describe the procedure and results of the experiments we performed aiming to quantify the terms $\delta_{rep}$, $\delta_{drift}$, $\delta_{temp}$ and $b$ of equation (2) as well as to assess the overall metrological performance of the sensor.

### *4.1. Bias and repeatability*

The smartphone was fastened on a metallic block with low geometric errors (flatness, perpendicularity and parallelism). The positioning of the equipment on the sinus table was realized with the aid of a precision square. This setup is presented in figure 2.

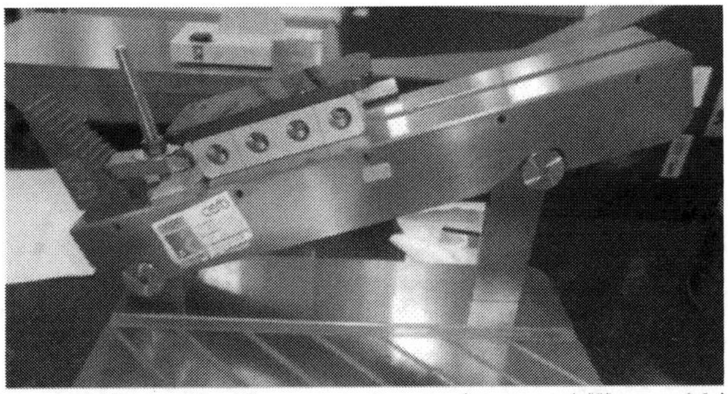

**Figure 2.** Setup for the assessment of repeatability and bias

For the evaluation of bias and repeatability, the smartphone was positioned in four different orientations with respect to the inclination generated by the sinus table, as shown in figure 3. This procedure allowed the realization of both positive and negative angles. Furthermore, the correction of the misalignment between the sensor and the sinus table was possible using the measurement data of these four positions. Three measurement cycles were performed in each position. Each measurement cycle was concluded in about 3 min.

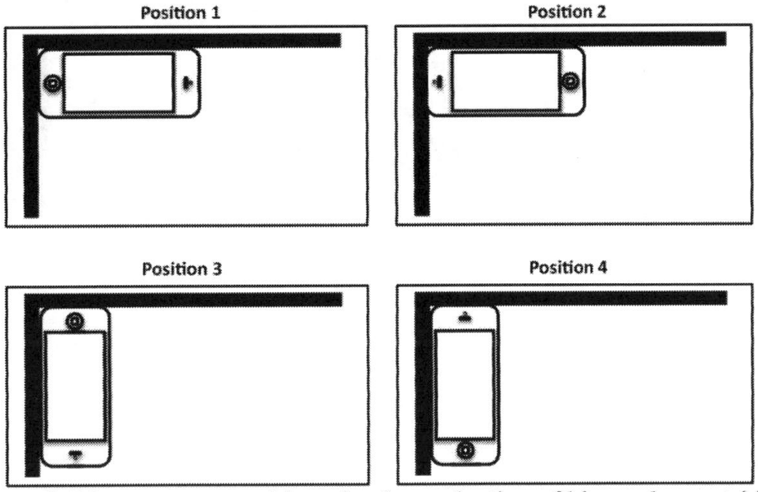


**Figure 3.** Measurement positions for the evaluation of bias and repeatability

The displacement of the sinus table was done smoothly in order to reduce the influence of dynamic effects.

Journal of Physics: Conference Series **648** (2015) 012007 doi:10.1088/1742-6596/648/1/012007

The repeatability was evaluated only for the points 0° and 20.49° at positions 1 and 3. At both points we made 20 consecutive observations, moving the table after each measurement.

The bias and repeatability experiments where performed in a temperature controlled metrology laboratory in which the temperature is kept in the range between 19.5°C and 20.5°C. For all measurements we adopted the average of 100 consecutive values as indication. These values were captured after stabilization of the signal.

The bias curves for the pitch and roll axes are displayed in figures 4 and 5. The dashed line represents the bias without correction of the misalignment of the sensor. The bias was calculated according to equation (3). $\overline{\theta_{ind.i}}$ is the average of the angles observed in the three measurement cycles for point *i*. $\theta_{ref.i}$ is the reference angle for point *i*, which results from the length of the gauge blocks as well as the distance between cylinders and the leveling height of the sinus table. $C_{align}$ is the correction coefficient for the misalignment between the sensor and the sinus table. This coefficient is calculated based on the measurement results at position 1 and 2 for the pitch axis according to equation (4) and positions 3 and 4 for the rolls axis according to equation (5). The maximum absolute bias are $/b/=0.14°$ for the pitch axis and $/b/=0.16°$ for the roll axis.

$$b_i = \overline{\theta_i} + \theta_{ref.i} \cdot \left(C_{align} - 1\right) \tag{3}$$

$$C_{align,pitch} = \frac{1}{2}\left[\left(\frac{\overline{\theta}_{pitch} - \theta_{ref,pitch}}{\overline{\overline{\theta}}_{roll}}\right)_{\theta=20,49;P1} + \left(\frac{\overline{\theta}_{pitch} - \theta_{ref,pitch}}{\overline{\overline{\theta}}_{roll}}\right)_{\theta=-20,49;P2}\right] \tag{4}$$

$$C_{align,roll} = \frac{1}{2}\left[\left(\frac{\overline{\theta}_{roll} - \theta_{ref,roll}}{\overline{\overline{\theta}}_{pitch}}\right)_{\theta=20,49;P3} + \left(\frac{\overline{\theta}_{roll} - \theta_{ref,roll}}{\overline{\overline{\theta}}_{pitch}}\right)_{\theta=-20,49;P4}\right] \tag{5}$$

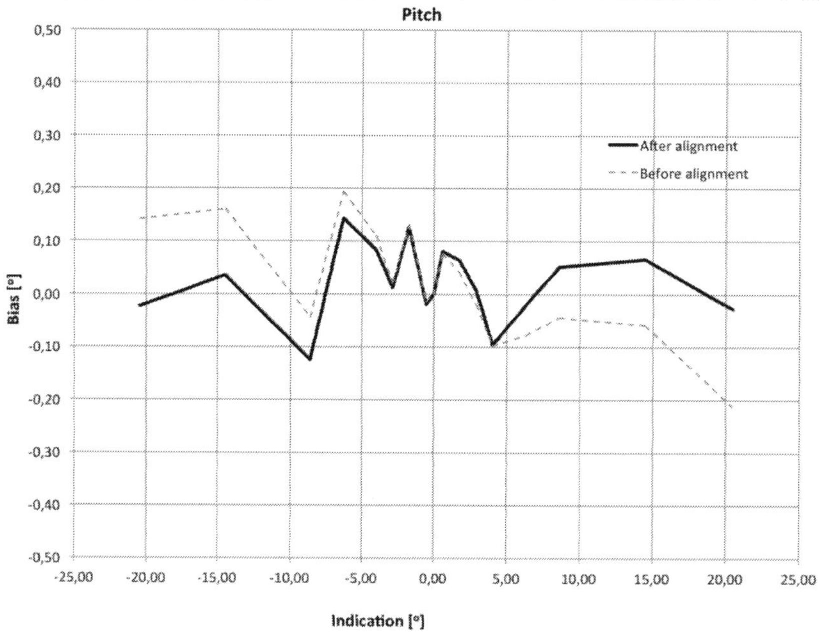


**Figure 4.** Bias curve - pitch axis

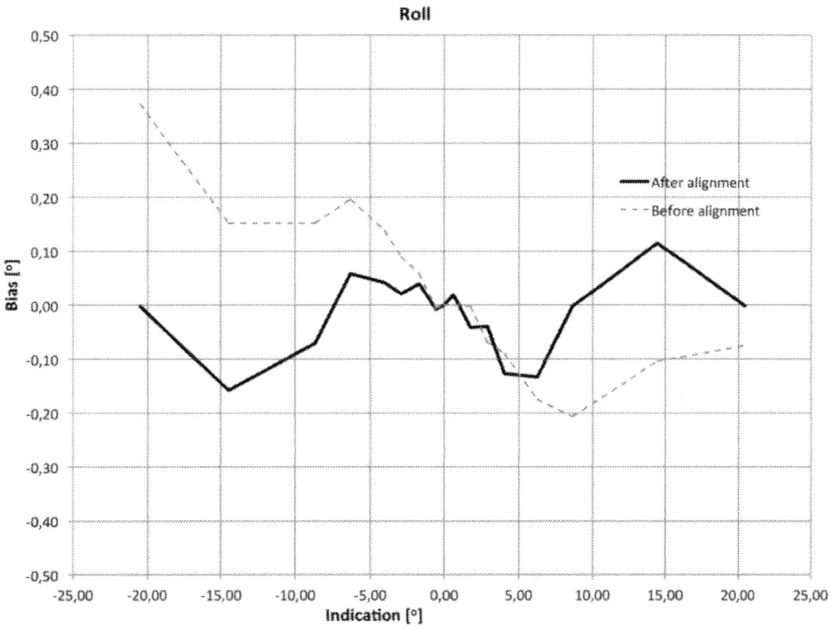


**Figure 5.** Bias curve - roll axis

The results for the repeatability can be found in table 1. The sample standard deviation $s$ was calculated based on the results of 20 consecutive observations.

**Table 1.** ***Repeatability***

| Axis | $\theta$ [°] | $s$ [°] |
|---|---|---|
| *Pitch* | 0,000 | 0,015 |
| | 20,487 | 0,034 |
| *Roll* | 0,000 | 0,016 |
| | 20,487 | 0,033 |

### *4.2. Signal drift*

As already mentioned, the raw output signal of a gyroscope is proportional to the rotation rate of the object. The angular displacement is obtained by integrating this signal. Consequently, a high accuracy measurement of angular displacement is only possible if the bias of the rotation rate signal is stable in time and thus can be corrected.

The bias instability of gyroscopes is typically evaluated by means of the Allan variance ($\sigma_a^2$) and its square root, the Allan deviation ($\sigma_a$) [1,14]. A long sequence of raw data (rotation rate) is used for this analysis. The sensor is kept as stable as possible during the collection of this data sequence. This sequence is then divided in $n$ bins of equal length. Each of these bins corresponds to an integration time $\tau$. The analysis starts with a small integration time, e.g. $\tau = 1/f_{aq}$, where $f_{aq}$ is the frequency of the data acquisition. The average of the data $y(\tau)_i$ is then calculated for each bin $i$. The integration time is successively incremented, always in a way that the number of bins is an integer. This procedure is

Journal of Physics: Conference Series **648** (2015) 012007 doi:10.1088/1742-6596/648/1/012007

repeated until nine bins are obtained. The Allan deviation $\sigma_a(\tau)$ is calculated for each integration time $\tau$ according to equation (6). The results for $\sigma_a(\tau)$ are plotted in a log-log diagram.

$$\sigma_a(\tau)=\sqrt{\frac{1}{2(n-1)}\sum_i\left(y(\tau)_{i+1}-y(\tau)_i\right)^2} \tag{6}$$

For short integration times, the Allan deviation is dominated by the signal noise. As the integration time increases, $\sigma_a(\tau)$ decreases due to the filtering effect. However, at some point $\sigma_a(\tau)$ starts to increase. This is related to the so-called rate random walk in the sensor, which is the inherent instability in the output of the sensor. The bias instability is defined at the minimum point of the $\sigma_a(\tau)$ curve [4].

We used a signal sequence of 10.5 h in our analysis. The raw signal for the pitch axis is show in figure 6. The result for the roll axis was similar. The standard deviation of the signal along the acquisition time was 0.25°/s for pitch and 0.26°/s for roll.

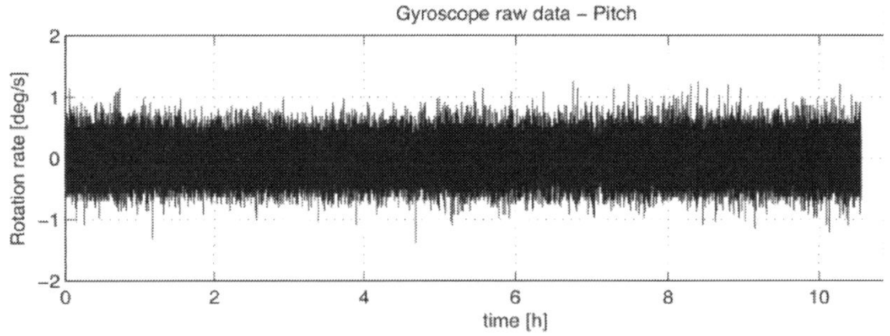


**Figure 6.** Raw signal - pitch

The Allan deviation curve for the pitch axis is shown in figure 7. The curve for the roll axis is similar. The minimum point of the curve (bias instability) was $9.0 \times 10^{-4}$°/s (3.2°/h) for $\tau = 3543$ s for the pitch axis and $10.0 \times 10^{-4}$°/s (3.7°/h) for $\tau = 3402$ s for the roll axis. However, it is not clear if the acquisition time we adopted was sufficient to reach the inflexion point of the curve.

Both the noise level of the signal and the bias instability were better than expected, compered to other MEMS based sensors. For instance, Aydemir et al. measured a bias instability of 41°/h and a signal standard deviation of 0.69°/s for a dedicated inertial measurement system [3]. Another dedicated MEMS sensor features a specification of 20°/h for the bias instability [15].

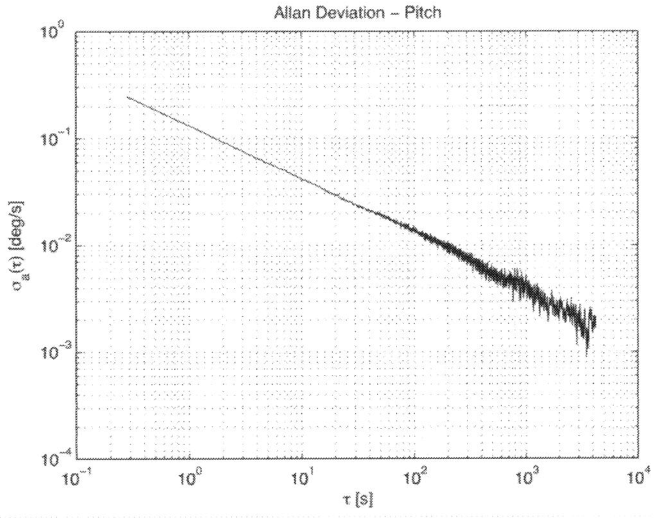


**Figure 7.** Allan deviation – pitch

Additionally, we evaluated the stability of the angle data provided by the sensor directly by monitoring this parameter over approximately 9.5 h, keeping the sensor stationary. The results are depicted in figure 8. An average drift of 0.036°/h was observed for the pitch axis and 0.008°/h for the roll axis. Although the level of noise is similar for both axes, the data of the roll axis presented randomly spaced peaks of around 0.1°. The cause of this effect could not be identified.

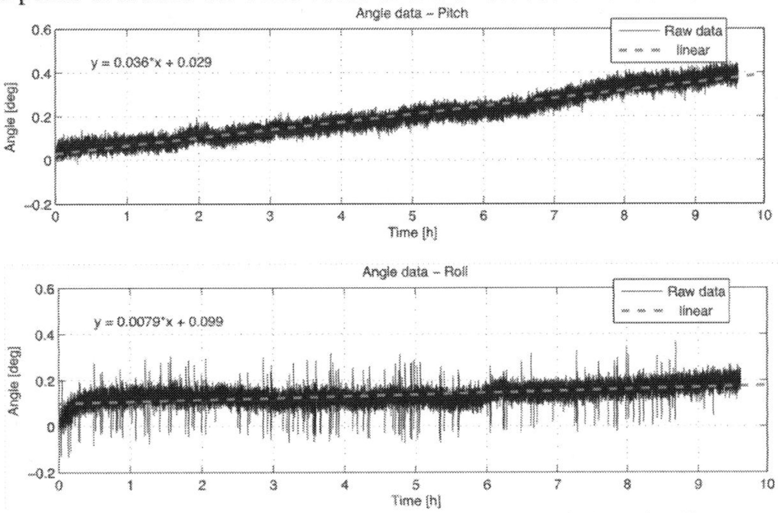


**Figure 8.** Stability of the angle data – pitch and roll

## *4.3. Influence of temperature variation*

Temperature variation due to changes in the environment and self-heating is reported to affect the stability of the rotation rate bias [1,8]. In this study we did not evaluate the influence of temperature variations under controlled conditions. However, since the average drift of the rotation rate signal was not significant in the data collected for the Allan deviation technique (see figure 6) we assume that the influence of temperature variations (at least those generated by self-heating) is negligible or internally compensated.

## 5. Measurement uncertainty budget

Since no significant influence of temperature variations could be identified, we simplified the model of equation (2), resulting in the final model of the measurement process (equation (7)).

$$\theta_{ind} = \theta + \delta_{rep} + \delta_{drift} + \delta_{ref} + b \tag{7}$$

The standard uncertainties $u$ related to this model with the respective degrees of freedom ν are estimated in sections 5.1 to 5.3. The sensitivity coefficient for all uncertainty sources is 1, as we are assuming that the model is linear and additive.

### *5.1. Standard uncertainty related to the natural repeatability of the sensor ($u_{rep}$)*

The standard uncertainty is equal to the maximum standard deviation obtained from the experiments described in section 4.1, divided by the square root of the number of observations ($n'$) expected to be performed in the regular use of the sensor. A Student's t-distribution with $\nu_{rep} = n\text{-}1=19$ is assumed ($n$ is the number of observations used to determine $s_{max}$). Assuming $n'=1$, $u_{rep}$ is given by equation (8).

$$u_{rep} = \frac{s_{\max}}{\sqrt{n'}} = 0.034^\circ \tag{8}$$

### *5.2. Standard uncertainty related to the signal drift ($u_{drift}$)*

The angle signal of a gyroscope will drift due to uncorrected bias and bias instability, calibration errors and due to noise in the rotation rate signal (angle random walk) [1]. Instead of taking each one of these causes separately into account, we opted to an empirical black box approach. The maximum observed average drift of the angle signal presented in figure 9 was used. This approach presents the advantage of being simple, but there is a risk of sub-estimating this source of uncertainty, since it takes only errors in account that are observed under static conditions.

A uniform distribution is assumed. The standard uncertainty is estimated according to equation (9), where $t$ is the time, in hours, in which the sensor is used without submitting it to a referencing procedure.

$$u_{drift} = \frac{drift_{\max} \cdot t}{\sqrt{3}} = (0.021 \cdot t)^\circ \tag{9}$$

### *5.3. Standard uncertainty related to the reference system ($u_{ref}$)*

The uncertainty of the angle generated by the sinus table depends on the uncertainty of the gauge blocks and the uncertainty of the geometric parameters of the sinus table – distance between cylinders and leveling height. The expanded uncertainty of the reference system, for a coverage probability of 95.45%, was estimated to be $U_{ref}$ = 0.007° (coverage probability of 95,45%, $k$ = 2). A normal distribution is assumed. The standard uncertainty is calculated according to equation (10).

$$u_{ref} = \frac{U_{ref}}{2} = 0{,}004^\circ \tag{10}$$

### *5.4. Combined standard uncertainty ($u_c$), effective degrees of freedom ($\nu_{eff}$), and expanded uncertainty ($U$)*

The combined standard uncertainty is given by equation (11).

$$\begin{aligned} u_c^2 &= u_{rep}^2 + u_{drift}^2 + u_{ref}^2 \\ u_c &= \left(\sqrt{0.0012 + 0.00044 \cdot t^2}\right)^\circ \end{aligned} \tag{11}$$

The effective degrees of freedom $\nu_{eff}$ are estimated by the Welch-Satterthwaite equation (equation (12)). There is a direct correlation between $\nu_{eff}$ and the measurement time $t$. We therefore used a relatively small $t$ of 0.1 h, which will result in a smaller $\nu_{eff}$, a higher coverage factor $k$ and a higher expanded uncertainty $U$.

$$\upsilon_{eff} = \frac{u_c^4}{\sum \frac{u_i^4}{\upsilon_i}} \qquad (12)$$

$$\upsilon_{eff} = 20$$

The expanded measurement uncertainty $U$ is given by equation (13). The coverage factor corresponding to $\nu = 20$ and a coverage probability of 95.45% corresponds to $k = 2.13$.

$$U = k \cdot u_c$$
$$U = 2.13 \cdot \left(\sqrt{0.0012 + 0.00044 \cdot t^2}\right)^{\circ} \qquad (13)$$

The maximum error is calculated according to equation 1, resulting in equation (14).

$$E_{\max} = U_{\max} + |b|_{\max}$$
$$E_{\max} = 2.13 \cdot \left(\sqrt{0.0012 + 0.00044 \cdot t^2}\right)^{\circ} + 0.16^{\circ} \qquad (14)$$

For a measurement time of $t$ = 1 h, the maximum error, rounded to one decimal place, is $E_{max} = 0.3^{\circ}$.

## 6. Conclusions

In this paper we evaluated the performance on angular displacement measurements of a MEMS gyroscope integrated in a smartphone. We performed experiments to quantify bias and repeatability of the sensor within the range of -20° and 20° for both pitch and roll axes. The bias stability of the raw rotation rate signal was evaluated by means of the Allan deviation. The stability of the angle signal was also evaluated through the acquisition and analysis of a long data sequence. The measurement uncertainty was estimated according to GUM.

We estimated a maximum error of 0.3°, for a measurement time of 1 h, which is significantly higher than the specification of a typical dedicated inclinometer (e.g. $E_{max} = 0.1^{\circ}$). However, we still consider this result positive, since the sensor integrated in the smartphone was not designed for accurate angular displacement measurement, but primarily for the estimation of the orientation of the smartphone in space. The bias stability and noise level of the raw rotation rate signal were better that expected, compared to dedicated MEMS-based sensors.

We could not investigate the influence of temperature variations under controlled conditions, due to the lack of an appropriate thermal chamber. We decided to neglect this influence in our measurement uncertainty budget, because no significant drift of the rotation rate signal was detected during an acquisition time of about 10.5 h. However, this hypothesis can only be confirmed by a controlled experiment.

It is important to mention that we performed the experiments under rather favorable measurement conditions (e.g. low temperature variation, low dynamic measurement, short measurement time). It is clear that the actual measurement error can be significantly higher depending on the measurement conditions at the actual application.

**Acknowledgement**

The authors thank CNPq for the financial support and CERTI Foundation for providing the facilities to conduct the experiments.

**References**

[1] Woodman O J 2007 An introduction to inertial navigation University of Cambrige *Computer Laboratory: Technical Report* **696**

[2] Schaeffer D 2012 MEMS Inertial Sensors: A Tutorial Overview *IEEE Communications Magazine*, **51** 4 100-109

[3] Aydemir G A and Saranli A 2012 Characterization and calibration of MEMS inertial sensors for state and parameter estimation applications *Measurement* **45** 1210-1225

[4] Nieminen T, Kangas J, Suuriniemi S and Kettunen L 2010 An enhanced multi-position calibration method for consumer-grade inertial measurement units applied and tested *Meas. Sci. Technol.* **21** 105204

[5] Zhang Y and Yang T 2012 Modeling and compensation of MEMS gyroscope output data based on support vector machine *Measurement* **45** 922-926

[6] Li Y, Niu X, Zhang Q et al. 2012 An in situ hand calibration method using a pseudo-observation scheme for low-end inertial measurement units *Meas. Sci. Technol.* **23** 105104

[7] Zhang H, Wu Y, Wu W et al. 2012 Improved multi-position calibration for inertial measurement units *Meas. Sci. Technol.* **21** 015107

[8] El-Diasty M, El-Rabbany A and Pagiatakis S 2007 Temperature variation effects on stochastic characteristics for low-cost MEMS-based inertial sensor error *Meas. Sci. Technol.* **18** 3321-3328

[9] Ilyas M, Yang Y, Qian Q S and Zhang R 2013 Low-cost IMU/odometer/GPS integrated navigation aided with two antennae heading measurement for land vehicle application *Proc. Control and Decision Conference (CCDC) (Guiyang, China, 2013)*

[10] Mitutoyo America Corporation 2014 Digital Protractor Series 950 -Pro 360 Pro 3600, http://ecatalog.mitutoyo.com/Digital-Protractor-Series-950-Pro-360-Pro-3600-C1438.aspx (accessed 03.2014)

[11] BIPM 2008 Evaluation of measurement data — Guide to the expression of uncertainty in measurement, BIPM, JCGM 100:2008.

[12] Bit Shape Software 2014 GyrOSC: OSC messanger for motion sensors, http://www.bitshapesoftware.com/instruments/gyrosc/, (accessed 03.2014)

[13] Processing Foundation 2014 Processing 2, http://www.processing.org, (accessed 03.2014)

[14] Stockwell W 2014 Bias stability measurement: Allan Variance Crossbow Technology Inc. <http://www.xbow.com/Literature/ Application_Notes_Papers/Gyro_Bias_Stability_Measurement_using_Allan_Variance.pdf> (accessed 03.2014)

[15] SBG Systems, IG-500A, http://www.sbg-systems.com/docs/IG-500A-Leaflet.pdf (accessed 03.2014)

[16] Park M 2004 Error analysis and stochastic modelling of MEMS based inertial sensors for land vehicle navigation systems Thesis: University of Calgary

# Vertical positioning surveillance by magnetostrictive transducer

**A S Carvalho, C Lopes, O O Pinto, C D Nunes, M F Borges and T R Strohaecker**
Laboratory of Physical Metallurgy (LAMEF), Federal University of Rio Grande do Sul (UFRGS), Av. Bento Gonçalves, 9500 (Building 104), Porto Alegre, RS, Brazil, ZIP 91509-900

E-mail: telmo@demet.ufrgs.br

**Abstract**. This work purpose is to create a positioning automated system of a tailstock to perform impact resistance tests on silicone mammary implants. This system is capable to measure and oversight the positioning through interrelation between three main components: programmable logic controller, human-machine interface and magnetostrictive transducer. Together, these components form an operational closed loop that ensures an appropriate positioning for the impact device. The paper describes how the closed loop works and also the algorithm implemented in the programmable logic controller which surveys the positioning. As a last topic, the paper presents the operator work on the machine's operation in conjunction with human-machine interface. The results were satisfactory and in accordance with the limits determined on ABNT NBR ISO 14607 for this method of tests.

## 1. Introduction

After resolution of the National Health Surveillance Agency (ANVISA), which establishes the minimum requirements for identity and quality for silicone mammary implants and imposes the requirement for certification of conformity of the product within the Brazilian System of Conformity Assessment (Sbac), laboratory analyses to verify the resistance of the prosthesis made itself necessary [1]. Hence it was designed a machine for testing the impact resistance of mammary implants based on the characteristics established in Annex E of [2].

There are some machines that perform impact resistance tests on silicone mammary implants in Brazilian laboratories, however none of them is as automated as the one shown in this paper. In order to build a machine that reduces operator's influence and so on enhances the repeatability of the test [3], it was developed the project of the Impact Resistance Test Machine.

The Impact Resistance Test Machine was fully developed by the Laboratory of Physical Metallurgy (LAMEF/UFRGS), from designing the project and its mechanisms until the automation system as a whole, including the positioning's algorithm. This part of the project consists in a closed-loop system that controls the position of the impact mass at the correct height, which is going to be detailed in the article body. The impact mass is equal to 4.4 kg and its diameter is equal to 250 mm, according to [1].

The operation center of this system is a programmable logic controller, through it occurs the surveillance position of the impact device and the control of all others devices. The measure of the position of the impact device is done by a magnetostrictive transducer [4]. The machine operator has to inform the prosthesis data and to enable the next stage after each concluded stage.

Journal of Physics: Conference Series **648** (2015) 012008 doi:10.1088/1742-6596/648/1/012008

It is expected to obtain a reliable vertical positioning for the impact mass, guaranteeing the accuracy of the test as well as the correct operation of the test machine.

## 2. Methodology

The Impact Resistance Test Machine in mammary implants is largely automated, the operator controls all stages of the process through a human-machine interface (HMI), Delta Electronics™ - DOP-B. The machine operator needs to inform the height and mass of the prosthesis onto HMI. The programmable logic controller (PLC), Delta Electronics™ - DVP-SA2, uses this information to perform the calculation of the height from which the impact mass is to be released against the prosthesis, and then a stepper motor, Applied Motion Products™ - HT34-487, is triggered by just rotating the spindle which moves a tailstock fitted with an electromagnet. The tailstock goes down to find the mass, so the electromagnet gets attached with the mass, then the couple moves vertically to a certain point set by calculation.

Once the impact mass reaches the height calculated for the test, the operator is informed about this condition and it is requested the positioning of the prosthesis at the test platform. After this, the machine operator can release the impact mass by HMI again, so that the electromagnet is switched off and the impact between impact mass and prosthesis occurs. It is prompted for the operator to remove the prosthesis of the test site and the process is terminated.

An image of the machine can be observed in figure 1.

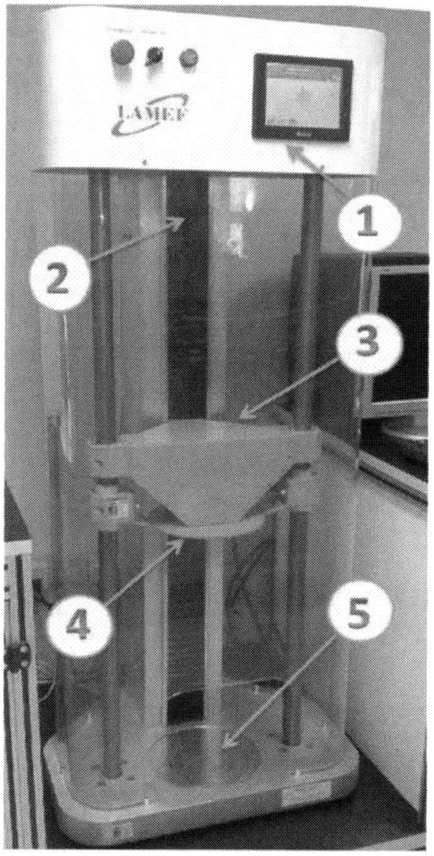


**Figure 1.** Impact Resistance Test Machine on mammary implants; 1. HMI; 2. spindle; 3. tailstock; 4. impact mass; 5. place of prosthesis (test platform).

### *2.1. Closed-loop system*

Positioning's routine of the impact mass at the height, where the test must be performed, is made jointly by an A/D module, Delta Electronics™ - DVP06XA-S, a PLC, and a linear displacement magnetostrictive transducer, GEFRAN™ – ONP1-A. These components form a closed loop system.

The transducer is positioned parallel to the spindle whereby the motor moves the tailstock. In this device there is a sensor which according with the position of the impact mass, returns a voltage level to the output of the transducer, acquired by A/D module. The PLC receives the signal already transformed into digital and compares this value with the value calculated according to equation (1).

Once the tailstock reaches the desired position, the PLC stops the stepper motor. Then, the machine operator is noticed that the test is ready to be performed.

In figure 2 it is shown a sketch of the closed loop representing this system.

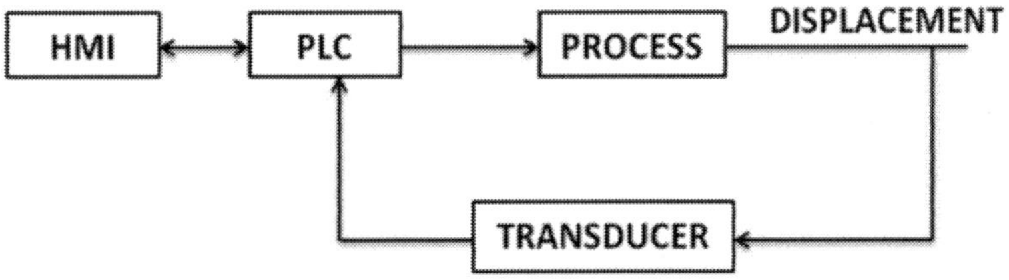


**Figure 2.** Closed-loop system

The sketch shows the loop feedback by the transducer. There is not a controller in the system, the PLC acts as a comparator, supervising when the impact device approximate the desired position. The PLC and HMI interrelation occurs with the exchanging of information by a serial communication. This way, the values inserted by the operator in HMI can be accessed by the PLC and the values calculated by the PLC can be accessed by the HMI.

### *2.2. Methodology of positioning the impact mass*

The machine operator measures the mass of the prosthesis on a precision weighing-machine, BEL Engineering™ - MARK M 2202, and observes the prosthesis height provided by the manufacturer. This data is entered via an HMI and accessed by PLC. The calculation of the distance where the impact mass should be in relation to the test platform, before the resistance impact test, is given by equation (1).

$$H[mm] = m[g] \cdot 0{,}95[mm/g] + h[mm] + 144[mm] \tag{1}$$

Where H[mm] is the distance between the test platform and the impact mass and m[g] and h[mm] are, respectively, the mass and height of the prosthesis.

With this calculated value, after the release of the operator, the PLC triggers three of its digital outputs: a pulsed signal in an input of the driver, Applied Motion Products™ - STR4/8, which drives the stepper motor, rotating the spindle in order to move the tailstock; another constant signal that drives the electromagnet in order to maintain fixed the impact mass to tailstock; and the last with a constant signal into another input of the driver, which defines the direction of motor rotation.

The driver that performs the communication between PLC and motor had its settings made under engine operation conditions. In this system the driver was configured with a ratio of 200:1 with microstep emulation, in which every 200 pulses received from the PLC on driver, representing a complete turn on the motor shaft. Thus, it ensures that the movement proceeds smoother [5], without compromising the structure of the machine.

Journal of Physics: Conference Series **648** (2015) 012008 doi:10.1088/1742-6596/648/1/012008

### *2.3. PLC – transducer communication*

The A/D module works with a resolution of 12 bits from -10 to +10V. The linear displacement magnetostrictive transducer returns in its output signal of 0.1 to +10.1V [6]. The use of these devices together provides a value to the PLC of 12 bits with sign bit.

In practice, the PLC quantizes this value received, from 0 to 2047, and transforms it into a value, in mm, indicating the distance between the impact mass and the test platform. The calculation performed by the PLC can be seen on equation (2).

$$H[mm] = -0{,}62[mm/bit] \cdot B[bit] + 1223{,}43[mm] \tag{2}$$

Where H[mm] is the height of the impact mass and B is the number of bits received by module A/D from the displacement transducer. This transducer was calibrated by a laboratory certified by the Brazilian Network Calibration (RBC).

### *2.4. PLC's algorithm*

The PLC program does the surveillance of the position of the tailstock with two conditions: if the tailstock is above or if it is below the desired position. The PLC uses the value obtained from equation (2) to compare with the value obtained from equation (1). The stepper motor stops when the value of equation (2) is in a range of ± 1 of value obtained by equation (1).

The PLC performs the comparison if the value arising from the transducer is above the calculated value, if it is positive, the stepper motor is triggered and spins the spindle making the tailstock going down. In case the result of the first comparison being negative, the PLC does a new comparison, this time to verify if the value arising from the transducer is below the calculated value, if it is positive, the stepper motor is triggered and spins the spindle moving the tailstock up. In case the result of the second comparison being negative, it means that the tailstock is already in reference position and then the positioning routine is finished.

A flowchart of routine process is shown in figure 3.

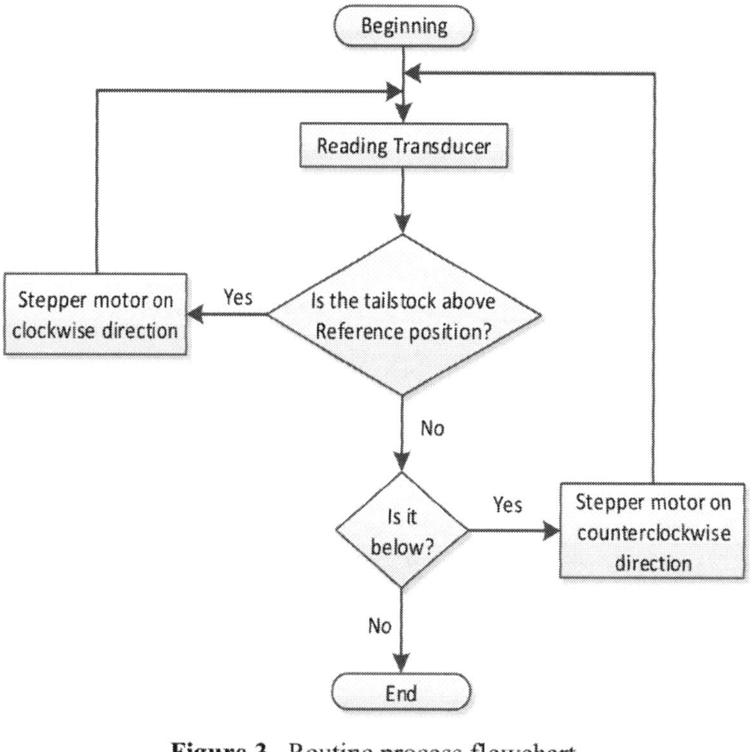


**Figure 3.** Routine process flowchart

*2.5. HMI – machine operator communication*

The operator's function in the Impact Resistance Test Machine on mammary implants is the insertion of data of the prosthesis, the enabling of every new stage and the placing and withdrawal of the prosthesis from the test platform. The communication between operator and machine is performed by a HMI. Through the HMI the operator can insert the mass and height of the prosthesis which are used in the calculation of the height that the impact mass is to be released in the test. It is from the HMI that the operator also can move onto a new stage of the process and can cancel the process at any time.

The operation of the machine is divided in some stages that demand the participation of the operator to occur:

In the first stage the operator enters the values of mass and height of the prosthesis to be tested in the HMI screen, as can be seen in figure 4, then he presses the 'OK' button to advance to the next stage. The values entered by the operator will be available to PLC use them in equation (1).

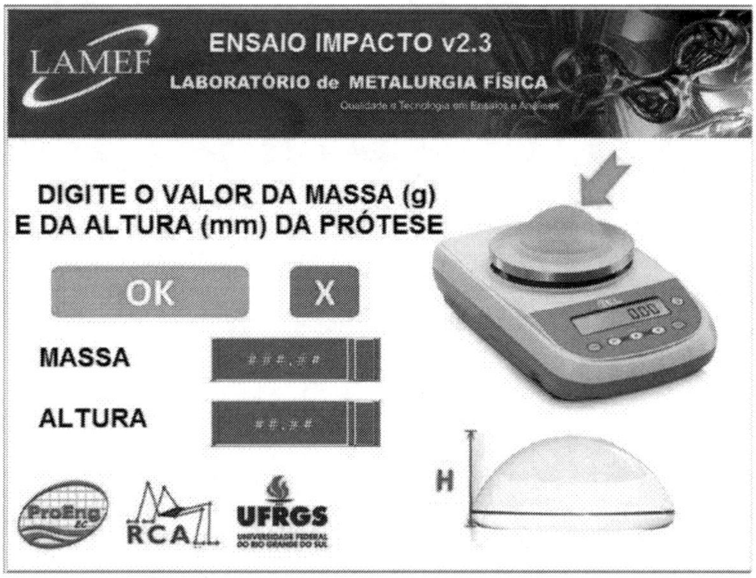


**Figure 4.** Inclusion of information screen

In the second stage occurs the positioning of the impact device through the closed loop system and algorithm already mentioned in the paper. The operator has the function to check if the data inserted is correct and print the screen to save the information about the test, as can be seen in figure 5. At the end of the positioning, the HMI informs to the operator that the stage was concluded and the 'OK' button is enabled to the operator, allowing him to move onto the next stages.

In the third stage of the process it is requested to the operator the placing of the prosthesis at the test platform. After the placement of the prosthesis, the operator must inform to HMI that the impact is ready to be performed by pressing the 'OK' button.

**Figure 5.** Positioning and assay's information screen

In the last stage occurs the impact between prosthesis and impact mass. After the impact, it is requested to the operator to remove the prosthesis from the test platform and the process is finished. Finishing the assay, the HMI returns to home screen, where a new assay can be started or the machine can be switched off.

## 3. Results

It was necessary to define a range of ±1mm of the acceptable value for ending the displacement of the tailstock due to oscillations provided by transducer's signal. However, the closed-loop system behaves satisfactorily, showing a fast response to the procedures required in this project.

By setting the configuration of the stepper motor driver to a relation of 200 steps/rev resulted into a better positioning of the impact device. Thus, it was possible to decrease the range of the acceptable value, from 1.5mm to 1mm, around the desired position to the end of the displacement of the tailstock.

After the final results were obtained, the machine was evaluated and accredited by Inmetro to perform assays.

## 4. Conclusion

The closed-loop system's methodology applied to oversight the positioning of the impact mass provided the reduction of external disturbances and greater stability for the correct positioning.

The friendly interface of machine provided a simple operation, easy to understand and without requiring an elaborate training to authorize someone to operate the machine.

Through automation of the test it was possible to obtain a great repeatability on the impact device's positioning, with small oscillations of values.

From the values obtained in the positioning of the system, it was concluded that the system operates satisfactorily with the purposes of this project and is able to perform impact tests on mammary implants with accuracy and in accordance with the requirements in the standard.

## Acknowledgements

We thank the Laboratory of Physical Metallurgy and the University of Rio Grande do Sul for providing us with the tools and structure necessary for this work.

Journal of Physics: Conference Series **648** (2015) 012008 doi:10.1088/1742-6596/648/1/012008

**References**

[1] Inmetro 2012 Portaria nº 162, Retrieved Mar 01, 2013 from http://www.inmetro.gov.br/legislacao/rtac/pdf/RTAC001804.pdf
[2] Associação Brasileira de Normas Técnicas (ABNT) 2013 Non-active surgical implants - Mammary implants – Particular Requirements *ABNT NBR ISO* **14607**
[3] BIOPULS 2008 Solutions for Biomedical Testing from http://instron-romania.com/media/533548158dd2d.pdf
[4] Calkins F T, Flatau A B and Dapino M J 2007 Overview of magnetostrictive sensor technology *J. Intellig. Mat. Syst. Struct.* **18** 1057-66
[5] Applied Motion Products 2012 STR4-8 Hardware Manual from http://www.applied-motion.com/sites/default/files/hardware-manuals/STR4-8_Hardware_Manual_920-0030G.pdf
[6] GEFRAN 2013 Datasheet ONP1-A from http://www.gefran.com/tr/download/3584/attachment/en

Journal of Physics: Conference Series **648** (2015) 012009 doi:10.1088/1742-6596/648/1/012009

# Machine for development impact tests in sports seats and similar

**R M Gonçalves**[1]
Rhodes S /A, Av. Rhodes, Cambuí, MG, 37600-000, Brazil.

E-mail:[1] rodrigo_ufsj@yahoo.com.br

**Abstract.** This paper describes the stages of development of a machine to perform impact tests in sport seats, seats for spectators and multiple seats. This includes reviews and recommendations for testing laboratories that have needs similar to the laboratory where unfolded this process.The machine was originally developed seeking to meet certain impact tests in accordance with the NBR15925 standards; 15878 and 16031. The process initially included the study of the rules and the election of the tests for which the machine could be developed and yet all reports and outcome of interaction with service providers and raw materials.For operating facility, it was necessary to set entirely the machine control, which included the concept of dialogue with operator, the design of the menu screens and the procedures for submission and registration of results. To ensure reliability in the process, the machine has been successfully calibrated according to the requirements of the Brazilian network of calibration.The criticism to this enterprise covers the technical and economic aspects involved and points out the main obstacles that were needed to overcome.

## 1. Introduction

This work originated from the chance that the test laboratory Rhodes [1] faced, in search of sporting seats certification for supply to football stadiums for the 2014 World Cup to be held in Brazil.

The big challenge was started from the study and thorough interpretation of the standards to be used. To go deeper in the interpretation of these standards, we participate with the Technical Committee CB-15 ABNT-Brazilian Association of technical standards. Through this study it was possible to reconcile the development of a same equipment with ability to perform impact tests and meet the requirements of standards NBR15925-plastic Furniture for sporting events [2], NBR15878-Seats for spectators [3] and NBR16031 – multiple Seats [4]. The equipment has earned the designation of MT-006-impact Testing machine.

This equipment underwent a severe assessment regarding the attendance of the requirements laid down by the impact of the standards tests. During and after construction, the non-conformities identified are analyzed with the objective of guaranteeing quality and reliability in the results of tests conducted there. The following sections describe various aspects of the development of this equipment, establish their potential, operating facility, types of tests and the economic aspects of the project and summary at the end, what was achieved for this experience.

## 2. Development of the machine

### *2.1 Basic and accessory Structures*

To meet the requirements of the tests proposed by NBR15925 standards, 15878 16031, equipment and passed initially by a dimensional analysis. Mechanical devices have been created to ensure the positioning and the application of loads on specific points of the specimens to be tested.

Accessories have been developed called the standards as "impact test device of the seat and pendulum of impact". They are used for the application of loads during the test. According to the rules, these accessories must meet the dimensional criteria for the specific mass and must be calibrated. Figures 1 and 2 show the impact devices.

**Figure 1.** a) Device to impact seat test.

**Figure 2.** b) Pendulum impact.

a) Device for seat impact test – circular body, with 200 mm diameter, separated from the impact plan by coil springs and free to move in a line perpendicular to the plane of the central area of impact. The body and the associated parties with the exception of the springs must be fitted with a mass of 17 kg ± 0, 1 kg and all appliances including the mass and the impact surface shall have a mass of 25 kg ± 0, 1 kg. See Figure 3.

b) Pendulum impact - cylindrical object having mass ± 6,5kg 0,07kg, supported by one pin through a steel tube 38mm ± 2mm diameter and 2mm wall thickness. The distance between the pin and the center of gravity of the pendulum must be 1 m. The pendulum has on its end a 50IRHD rubber with a radius of 67mm. See Figures 4 and 5.

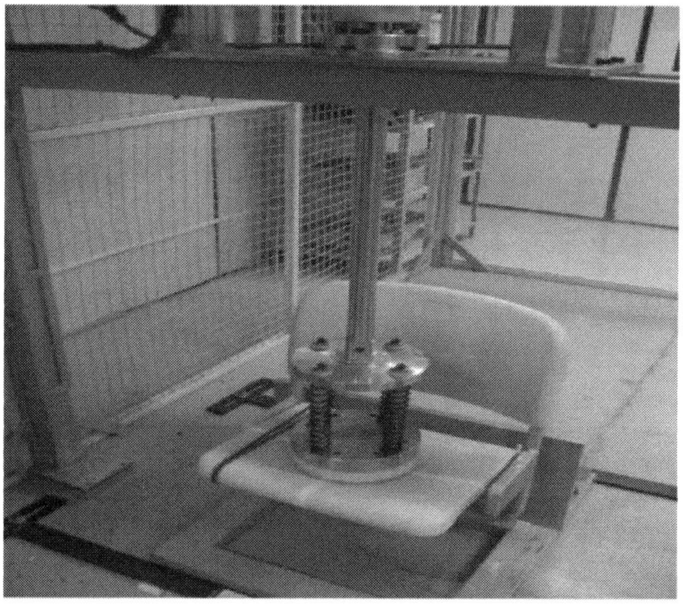

**Figure 3.** Impact on the seat.

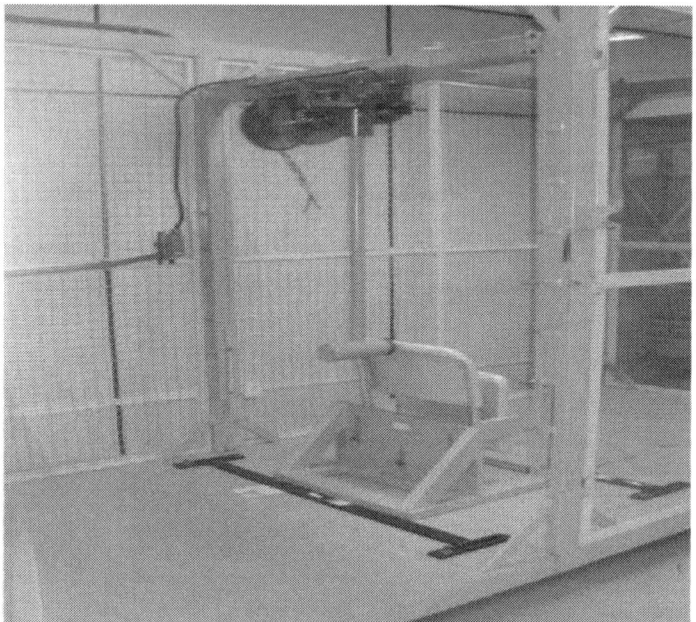

**Figure 4**. Impact on the backrest.

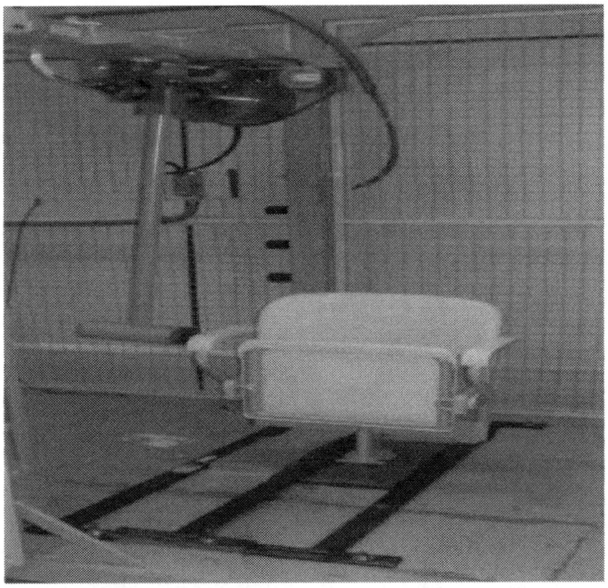

**Figure 5.** Impact on the arm rests.

### *2.2 Electropneumatic System*

The electro pneumatic command of this equipment was achieved through joint work with suppliers of equipment for industrial automation. Automatron Industrial Technology companies [5] and Festo Automation Ltda. [6] had important performance in this step.

The equipment is controlled by a PLC-Programmable logic controller, see Figure 6. The operation is semi-automatic equipment, must be initialized via operator. Data such as number of cycles, height of drop, drop angle of impact device and pendulum, can be viewed through the HMI-Human machine Interface and can be parameterized and initialized according to the types of tests and requirements, this configuration can be seen in Figure 7.

**Figure 6.** PLC-programmable logic controller.

**Figure 7.** HMI-Human machine interface.

In order to use the device to impact test of the seat, there is a valve for pneumatic actuation of control ascent and descent of two-cylinder actuators. With the aid of electromagnet, this system allows you to trigger the device up and down. The course or free fall height of the impact device on the seat are capable of programming and controlled by means of sensors, see Figure 8.
The pendulum of impact is driven by a stepper motor coupled to a clutch by means of two gears and transmission chain, as shown in Figure 9. This setting allows the command of appropriate tilt angle and free fall of the pendulum, in addition to meeting the values established by the standards.

**Figure 8**. Cylinder actuators + electromagnet (impact of seat).

**Figure 9.**Step motor + clutch (pendulum).

The figures 10 and 11 allow better understand the design philosophy that guided this intervention, and the results that were obtained here. Figure 10 highlights the buttonhole that is provided with the Programmable Logic Control which was adopted. As you can see, it includes, in a certain standard format, texture and color, a conventional numeric keypad to ten digits, data entry keys and return, and directional navigation keys to two orthogonal axes. According to a consistent pattern, but in a different color, there is another field with six function keys. The held programming can be seen in the PLC screen that allows you to track the entry of all data provided.

**Figure 10**. Keypad of programmable logic controller

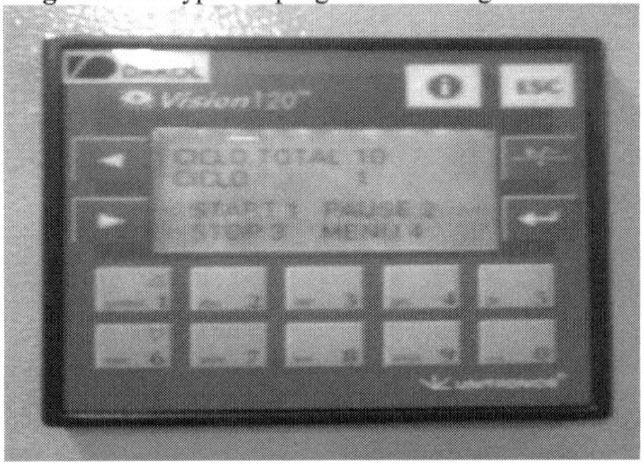


**Figure 11.** Screen programming PLC.

This command arrangement complies with current regulations in the literature, understood in Ergonomics, based on principles that seek the efficiency and safety in operation, and the minimum wear of the operator [7]. Correspondingly, the construction of the dialogue to be established between the machine and the operator as well as their presentation on the screen, should be based on ergonomic principles, which are also regulated by international technical standards [8, 9]. The names used for the measuring, their acronyms, and the sequence of presentation of grandeur were chosen in order to optimize the operation. With the same goal, were exploited the presentation graphics capabilities of the system screens offered.
As well as in Figure 11, the remaining arrangements for the presentation of Visual information always were prepared in order to provide clarity, ease of interpretation and cognitive reference comfortable. An evaluation of these results, made by skilled technician, indicated that they are perfectly in line with the recommendations of contemporary literature, and even more robustly established principles, such as those of the Gestalt theory [10].

## 3. Technical and operational data of test equipment

The following is a description of the technical characteristics of the machine after built for purposes of evaluation of the results of the project:
This is an outfit usually called impact Testing machine, dedicated to tests of sports seats, seats for spectators, multiple seats and their similar ones. Is a semi-automatic equipment, with supply voltage of 220Vac, 60 Hz frequency, power consumption 120W. She is endowed with a processor "Dakol V12022T38" (registered trademarks here eventually cited are the property of their legal owners).
The testing machine was calibrated as to its characteristics liable to calibration. The certificate was issued by [11] calibration laboratory accredited by the general coordination of Accreditation which is the Brazilian Government authority that provides accreditation in accordance with norm NBR ISO/IEC17025-General requirements for the competence of testing and calibration laboratories [12].
The equipment is capable of performing the tests whose descriptions follow below. These descriptions were transcribed or summarized technical norms NBR 15925, 15878 and 16031 already mentioned.
Impact tests of seat:
Are tests that evaluate the ability of the body-of-proof resist the freefall of impact of 25 ± 0, 1 kg at a height of 300 mm on the seat for 10 cycles.
Impact tests on the backrest:
Are tests that evaluate the ability of the body-of-proof resist the free fall of the pendulum of impact of 6, 5 kg ± 0, 07kg at a height of 620 mm or 68° angle on the outer rear edge of the backrest for 10 cycles.

An important feature of the system logic is highlighted below. The tests can be selected in programmed screen by pressing F4 = MENU, which belongs to the human machine interface already mentioned earlier. However, they cannot be changed during execution, once the program is inserted into the internal memory of the PLC (Programmable logic controller) is password protected. Therefore the operator has access only to the functions of start, finish and pause the test. This feature addresses explicitly the requirements of NBR ISO/IEC 17025 already mentioned in the chapter in which this standard deals with the integrity of data collected or handled by computerized systems. Generally, this feature helps ensure reliability test results, and this is an aspect very carefully observed in the process of accreditation of a laboratory.

## 4. Economic aspects of the project

Beyond the purely technical aspects regarding the report of the development of equipment in order to meet the requirements of the standards, that treated all the previous text, fits make a brief comparison of costs between the scenarios of this development. This analysis is justified, since this whole process if entered in the company's strategic plan Rhodes, manufacturer of components and accessories for the furniture industry, which has, among its goals, the permanent search for the quality. This search has indicated, on the one hand, the desirability of providing the company of an accredited testing laboratory as competent. On the other hand, it was necessary to harmonize this proposal with the availabilities and the company's expectations regarding the economic aspect of the investment.

The numbers shown below allow you to establish comparisons on which to draw a conclusion as to the relevance of the business.

Impact Testing machine:

Enterprise value = $ 48,363 .00

Ability of testing-types = 3 essays

Total value of a battery of 3 external laboratories tests = $ 6,000 .00

Number of tests that have already been carried out in a period of 12 months = 15 essays.

As for the value of the equipment, in a period of one year of tests the numbers indicate a reasonable estimate: the total cost of the project was $ 48,363 .00, therefore, in terms of equity, the balance was positive with return on investment at about $ 41,637 .00 or 86% of the amount invested.

As for the ability to perform tests: the machine is extremely versatile, meets the tests of three technical standards of following furniture. In addition, when it is obtained the accreditation of the laboratory for these tests by the Government authority, the lab will expand greatly its ability to provide compliance testing services to third parties, which may constitute a new source of revenue to the company.

## 5. Conclusion

This venture has been subjected to a comprehensive initial assessment, focused on basal issue: the need to impose certain rules of procedure in a productive activity, and the results observed as a result of this enforcement [13]. This evaluation indicated that a favorable balance was obtained.

The construction and development of impact test equipment met the requirements established by the rules NBR15925, 15878 and 16031. In addition, allowed the test laboratory Rhodes ease of interrelation with operator trials machine, calibration facility for issuance of the certificate RBC-Brazilian network of calibration.

As previously demonstrated, the project achieved a good result, in terms of return on investment.

Experience has shown the convenience of close cooperation between technical staff from both parties: providers and consumers of laboratory equipment. Also showed that there were gains for both parties, when company executives who participated in the study, as well as review and preparation of technical standards.

Noting also that this work was paramount to meeting deadlines agreed with internal and external clients, reduced cost of tests externally and mainly to guarantee Laboratory's competence in conveying reliability and quality in the results of the tests, giving an essential step to its accreditation by the Government authority, and the resulting benefits which may be derived.

**References**

[1] RHODES, www.rhodes.ind.br, Rhodes, Cambuí, Minas Gerais, 2013.
[2] ABNT NBR, 15925:2011- Móveis – Assentos plásticos para eventos esportivos.
[3] ABNT NBR 15878:2011- Moveis – Assentos para espectadores- Requisitos e métodos de ensaios para resistência e a durabilidade.
[4] ABNT NBR 16031:2012 – Moveis – Assentos múltiplos –Requisitos e métodos para resistência e durabilidade.
[5] AUTOMATRON, www.automatron.com.br, Automatron Industrial Technology, Cambuí, Minas Gerais, 2013.
[6] FESTO, www.festo.com.br Automation, Festo Ltda., São Paulo, SP, 2013.
[7] J.Dul and b. Weerdmeester, ergonomics Practice, sec. 3, ed. Edgard Blücher, São Paulo, SP, 1991.
[8] ISO, ISO 9241-10-Ergonomic requirements for office work with visual display terminals (VDTs)-Part 10: Dialogue principles, International Organization for Standardization, Geneva, 1996.
[9] ISO, ISO 9241-12-Ergonomic requirements for office work with visual display terminals (VDTs)-Part 12: Presentation of information, International Organization for Standardization, Geneva, 1998.
[10] I. Iida, ergonomics: design and production, sec. 10, ed. Edgard Blücher, São Paulo, SP, 2005.
[11] EXCEL METRO, www.excelmetro.com.br, Excelmetro in Metrology Services Ltd., Varginha, MG, 2013.
[12] ABNT NBR ISO/IEC 17025: 2005- .Requisitos gerais para a competência de laboratórios de ensaio e calibração.
[13] C. C. Bucich, rule enforcement in industrial production: Ergonomics in the processes of standardisation and certification, doctoral thesis, Universidade Federal do Rio de Janeiro, Rio de Janeiro, 2003.

# Laser Triangulation Profilometer for Inner Surface Inspection of 100 millimeters (4") Nominal Diameter

**P Buschinelli[1], T Pinto[1], F Silva[1], J Santos[2] and A Albertazzi[1]**
[1] Federal University of Santa Catarina (UFSC), Florianópolis - SC, Brazil
[2] CENPES, Petrobras, Rio de Janeiro - RJ, Brazil

E-mail: pvb@labmetro.ufsc.br

**Abstract.** A new axis-symmetrical system was developed for inner surface pipe measurement focused in the Oil & Gas industry, aiming the quantification of corrosion and others inner surface anomalies. The measurement principle, calibration, analysis of the measurement errors and laboratory and field results are described.

## 1. Introduction

Inner pipe inspection is of high interest especially for sewer lines and Oil & Gas (O&G) industry [1] [2] [3]. The inner pipe surface is generally measured to detect anomalies like corrosion and deformations. For the inspection of pipes in the O&G industry, sophisticated equipment, named Pipe Inspection Gauge or PIGs, are used. The most common measurement principles used by PIGs are magnetic flux leakage (MFL) [4], ultrasound [5] and mechanical probing [6]. Those devices are very robust, can measure several kilometers of pipes and usually are pushed by the own pipe-line oil or gas flow. PIGs are ideal for long-range pipe measurements, but they are not able to obtain very detailed inner surface information. On the other hand, optical systems are ideal when high density and accurate measurements are needed. Although, the pipe must be empty and clean or filled with a homogenous and transparent fluid.

Inner pipe laser inspection using a conical mirror was described by Inari; et al. (1994) [7]. A laser emitter and a conical mirror are used to generate a conical light sheet for measurement of small heat exchanger pipes (inner diameter range of 14.5 – 25.0 mm). Other systems using similar approaches were described in [8] [9] [10].

The group of Yoshizawa; *et al* (2006, 2007, 2012) [11] [12] [13] and ours (2007, 2012) [14] [15] independently developed an approach to measure the inner geometry of pipes from a single image using the so called "ring beam" or "radial light sheet" laser system.

In previous works we have described a system for the inspection of 152 mm (6") nominal diameter pipes [16] [15]. In the present paper, we describe the measurement principle, calibration, analysis of the measurement errors and laboratory and field results of a new system[3] for smaller (100 mm) diameter pipes.

[3] From now on, the laser profilometers for inner measurement of pipes of nominal diameter of 100 mm and 152 mm will be cited on text, respectively, as 4-inch profilometer and 6-inch profilometer.

Journal of Physics: Conference Series **648** (2015) 012010 doi:10.1088/1742-6596/648/1/012010

## 2. Optical Measuring Principle

In order to measure a full (360°) inner pipe section per acquired image, our system uses a 45° (half angle) conical mirror to project a laser plane perpendicular to the cone axis. The laser source must be tightly aligned to the tip of the mirror. A camera looks at the resulting laser line formed by the intersection of the light plane and the inner pipe wall. The system optical configuration is shown in figure 1. If everything is well aligned and the pipe section is circular, the resulting image viewed by the camera will be a circle.

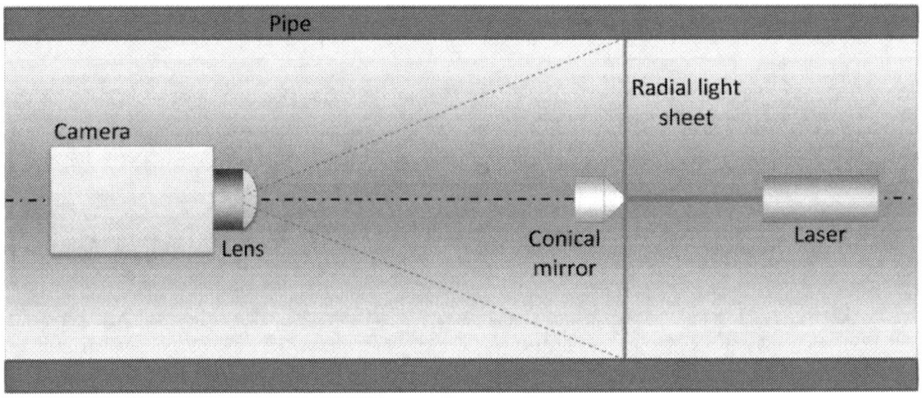


**Figure 1.** Measurement principle of the optical profilometer [15].

### 1.1. Image processing

To extract the laser line coordinates the image is analyzed in polar coordinates (see left image at figure 2) and discretized in angle steps $\Delta\theta$. A threshold filter is applied and the center position of the laser line is calculated (center of mass approach) for each discretized line. For our system, the image is discretized in 1440 angles (0.25° angular resolution) and the center is defined as the center of the image. The image processing result is an array of radius indexed by its angle. In order to reduce the processing time, a region of interest is delimited by a minimum and maximum searching radius. Also, the image is processed in a graphic card processor using CUDA parallel computing technique. The typical processing time in GPU is about 2 ms for the whole image (time measured in a Laptop with a GTX 285m graphic board). More details of the image processing can be found at [16].

The measurement is referenced in cylindrical coordinates, as shown in figure 2. For each acquired image, the actual system axial position (*Z* axis) and rotation (around *Z* axis) are measured by an odometer and an inclinometer, respectively.

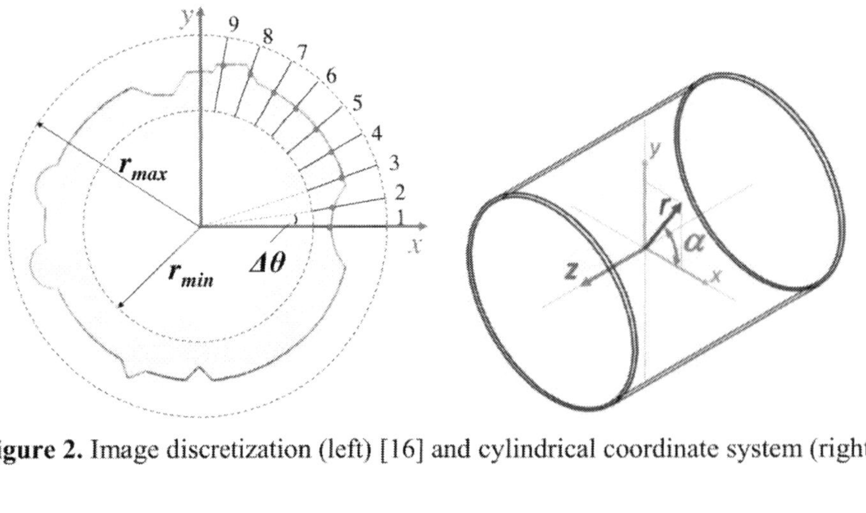


**Figure 2.** Image discretization (left) [16] and cylindrical coordinate system (right).

### *1.2. Optical Calibration*

There is a relationship between the radius in pixel of the laser line image and the actual radius of the pipe in real world coordinates. Ideally, this relationship would be linear, and equal for all angles. Although, due to imperfections, there are significant non-idealities, which can be divide in two main contribution groups: 1) the wide lens distortion plus the refraction on the acrylic wall results in a *nonlinear radius relationship*; and 2) the misalignment and imperfections of the components results in an *uneven behavior for different angles*. Beside those effects, the laser line viewed by the camera is not homogeneous, due to laser speckle. This effect is the main responsible for random errors of the system, and cannot be compensated during calibration. As a result, any laser line position is obtained with a ± 0.5 pixel repeatability.

In order to calibrate the system, we acquire images of an inner stepped cylindrical gauge (SCG) with rings diameters previously calibrated in a coordinate measuring machine (CMM). The SCG has 28 different cylindrical regions with nominal diameter steps of 4 mm. The profilometer is able to measure from the 2$^{nd}$ to the 13$^{th}$ ring, with respective calibrated diameter of 87.897 mm to 131.896 mm. The obtained mean radius in pixel are 327 and 509, which define the profilometer measuring range. Although, those values can be extrapolated a bit (about 10 pixels). The resulting mean resolution is of 0.121 mm/pixel (sensitivity of 8.27 pixel/mm).

Figure 3 shows the system measuring one of the rings of the SCG. The SCG is assembled on our calibration bench and carefully aligned with the profilometer. The bench allows a controlled displacement of the SCG (X, Y and Z axis) using linear translation stages. By moving the stepped cylindrical gauge along the system axis (Z axis) it is possible to submit the profilometer to different and well-known diameters. A third degree polynomial is fitted to relate image radius to ring radius. A different polynomial is fitted for each discretized angle.

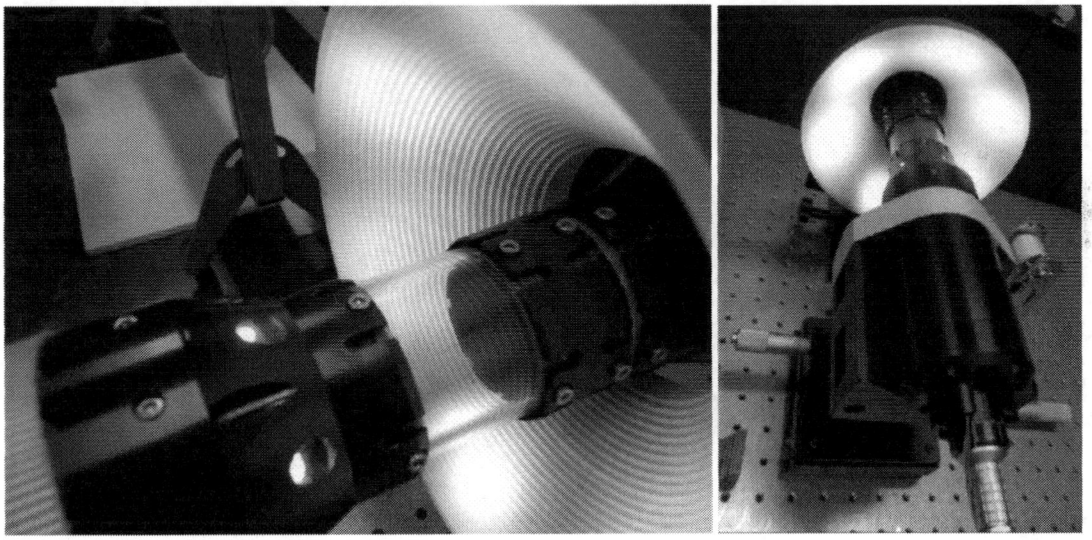

**Figure 3**. Profilometer on the calibration bench. Laser is illuminating one of the rings of the stepped cylindrical gauge. LEDs are turned on just for demonstration.

## 3. Prototype

The 4-inch profilometer uses a gigabit Ethernet (GigE) CCD camera capable of acquiring images of 1024x1024 pixels at a frequency of 60 Hz (frames per second). The square shape of the sensor is appropriate for the system, since its measuring range is circular and extra borders would be of no use. The theoretical system maximum speed is limited by the camera maximum frequency. A photo of the 4-inch profilometer is shown in figure 4. The laser source (660 nm; 30 mW) is powered through an enameled AWG 40 (~ 0.1 mm diameter) cooper wire (see figure 5). As it can be seen in figure 6, the wires did not affect the measurement (right image) and can be barely noticed on a normal camera image (left image). The previously developed profilometer [15] uses batteries to power the laser, thus, no wires pass through the field of view. The wired version make possible remote laser on/off switch, control its power and enable size reduction.

The 4-inch profilometer has a minimum passage diameter of 85 mm (3.3"); a diameter measuring range of 88 to 132 mm (3.5" to 5.2"); total length of 605 mm and mass of 3 kg.

To measure the displacement along the pipe, a mini optical encoder (1440 pulses per revolution) is attached to one of the prototypes wheels. Also, an accelerometer is used as inclinometer to measure and compensate undesirable rotation along the pipe axis. It must be noticed that the inclinometer will not operate correctly in pipes with axis close to vertical (aligned to the gravity).

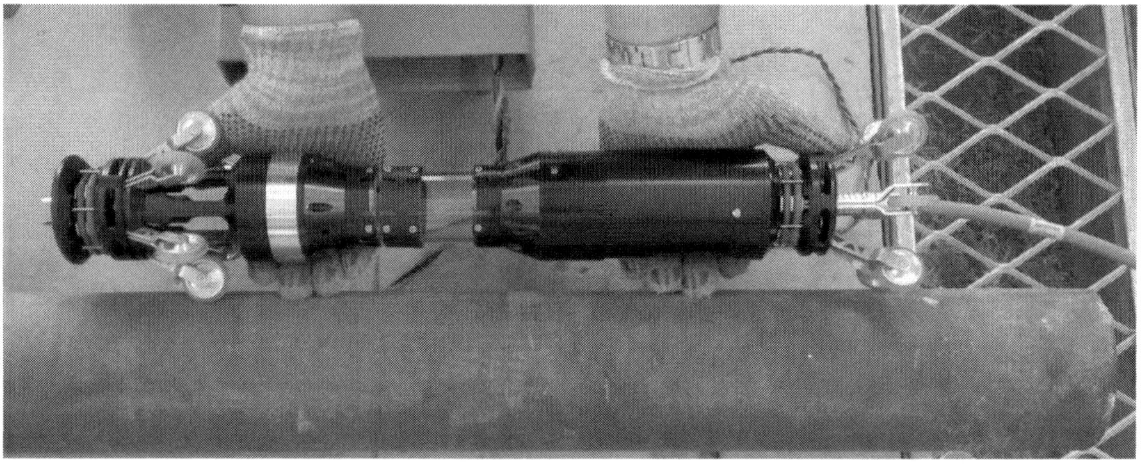

**Figure 4.** Photo of the 4-inch profilometer before insertion into a refinery pipe for inspection.

**Figure 5.** Detailed view of the AWG 40 wires for powering the laser.

A microcontroller counts the wheel encoder pulses and triggers the camera acquisition for each *p* number of pulses. Since it is almost impossible to maintain the system displacement at an enough constant speed, the displacement triggered acquisition keeps the Z spacing constant even with displacement speed variations. However, there is a maximum speed that cannot be exceeded, which could result in a frame rate request faster than the camera can handle, resulting in a section lost. In order to avoid that, we evaluated that a considerable safe speed to operate is limited by the acquisition frequency of about 45 Hz (75% of the maximum). Since different axial resolutions can be configured, the displacement speed can be calculated by multiplying the desired resolution (*dz*) by the safe maximum frequency (*dz**45). In our experiments, we use a nominal resolution of 1 mm, which results in 45 mm/s. The real displacement resolution is a multiple (number of pulses) of the odometer resolution (millimeters per pulse), achieved by calibration.

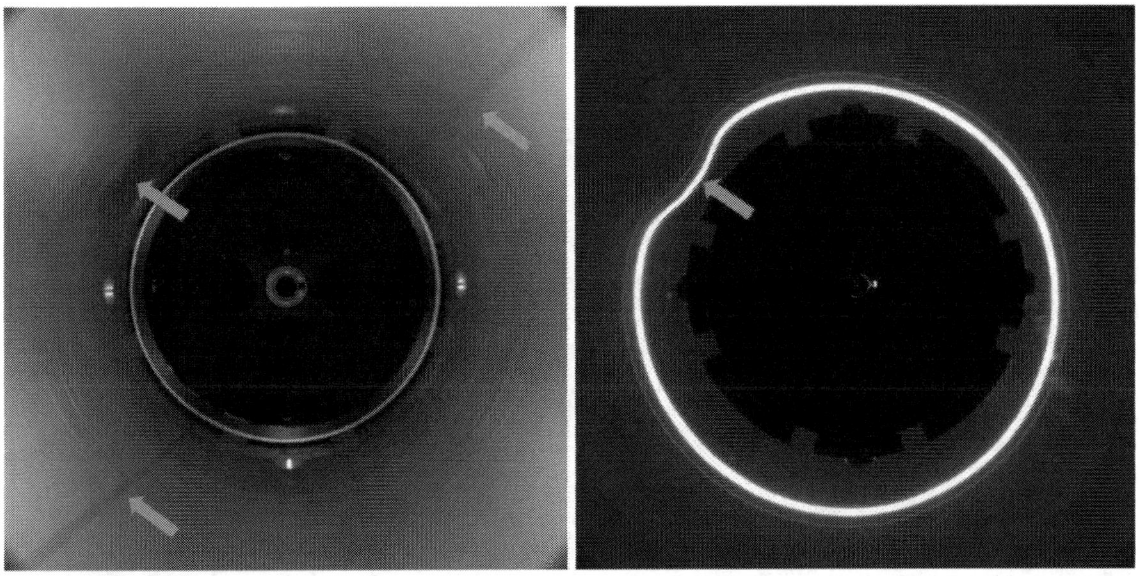

**Figure 6.** Photos taken by the 4-inch profilometer inside a plastic tube with the LEDs turned on (left) and with the laser turned on (right). The arrows points to the laser power wires (left) and to a pipe heat created imperfection (left and right).

The odometer calibration was done using a three meter reference straight guide and a laser interferometer with negligible uncertainty. The number of pulses obtained for a known distance of about 3 meters was acquired five times. The obtained mean relation in pulses per millimeters is of 0.072600 (sensitivity of about 13.7 pulses per millimeter), with a standard deviation of 7.2E-06. After calibration, the system was displaced several times by a set of known distances. The resulting error curve is displayed in the graph of figure 7. It can be noticed that the error increases with the distance, probably due to drift.

Imperfection on the pipe wall will affect the displacement measurement, since variations in the pipe radius will result in a larger displacement measurement than the actual displacement in the pipe axis. Changes in the odometer wheel diameter (due to abrasion or due to adhesion of material) will also increase the error, e.g.: a variation in the odometer wheel diameter of 0.1 mm results in an error increase of 0.3%.

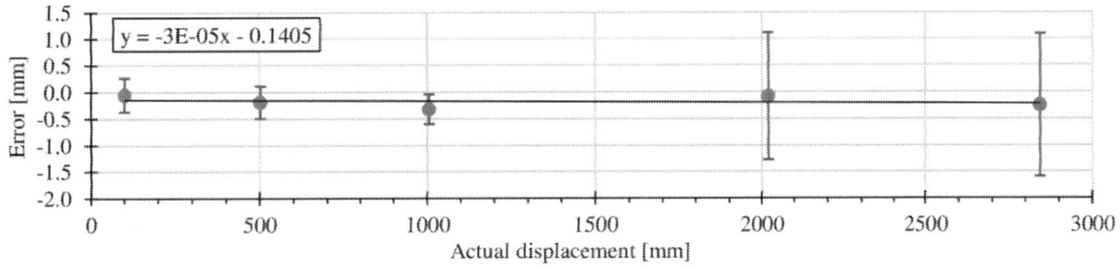


**Figure 7.** Odometer error curve for five different ranges. Mean error (points), confidence range and fitted line using least squares method (LSM). Vertical and horizontal axis represents, respectively, the error and the distance. Values in millimeters.

## 4. Results

The profilometer was tested and evaluated in laboratory and in onsite conditions. These experiments are described next.

*4.1. Optical Calibration Evaluation*
Two different procedures were used to take the calibration images: A) the SCG was aligned to ring 8$^{th}$; B) the SCG was re-aligned for each ring.

To align the SCG, it is moved until a fitted circle center X and Y coordinates are less or equal 0.03 pixels apart of the processing polar coordinate center. The circle fitting is done using least absolute deviation (LAD) method, which is more robust to outliers then the least square method [17].

To reduce the speckle random error, a median and a mean filter were applied to the radii data of all calibrations. The filter size was set to 11 elements and it was repeated 10 times (11x10). Figure 8 shows radius data obtained for the 8$^{th}$ ring before and after filtering. We observed that taking more than one image per ring did not improve the calibration, because the difference in radius in pixels is very low and mainly due to speckle effects.

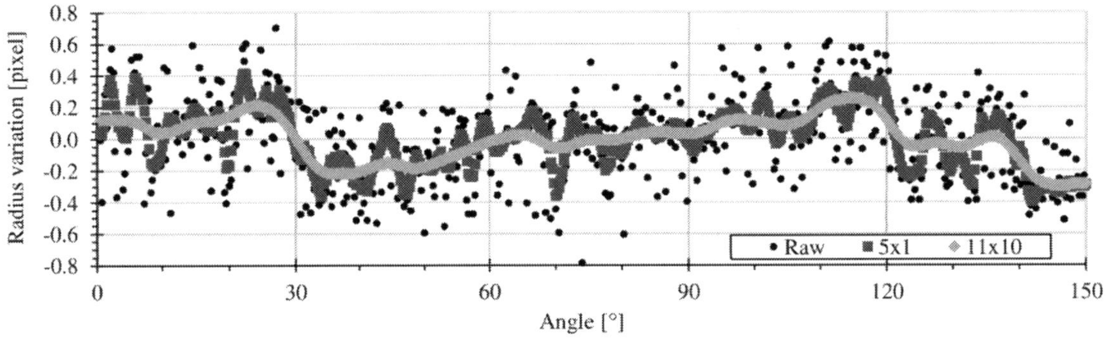


**Figure 8.** Radius data (raw and after filtered using a mean and median filter of *e* elements and applied *t* times *exj*). Data obtained from an image of the 8$^{th}$ ring.

*4.1.1 Polynomials Comparison.* The mean radii for each image used during calibration 'A' were calculated. Those mean values were used as input of the calibration polynomials. The plot result is shown in figure 9. Only part of the 1440 polynomials are plotted (for every 30°). As can be noticed, the resulting polynomials for different angles are almost close to straight lines and not very much different to each other.

To evaluate the differences between the obtained polynomials, the resultant radii in millimeters were subtracted from the obtained with the mean polynomial. The result is shown at figure 10. As one can see, for the same radius in pixel, there are a difference of about ±0.3 mm. This can be due to two main reasons: 1) uneven behavior for different angles; 2) misalignment in the trajectory of the SCG during calibration. It is impossible to tell which one is the main influence just by looking to the profilometer and calibration bench assembly. Although, if the *uneven behavior for different angles* is the main factor, the calibration 'A' (alignment only in one ring) must have a smaller measurement error. On the other hand, if the *misalignment in the trajectory of the SCG* is the main factor, calibration 'B' (realignment in all rings) must have a smaller measurement error.

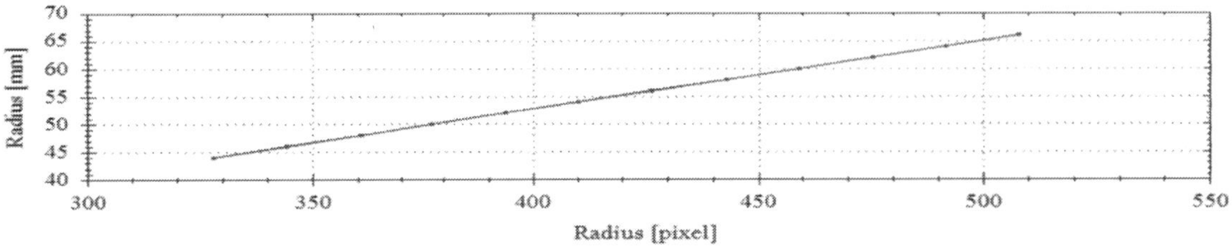


**Figure 9.** Calibration using method 'A' (alignment in only one ring) polynomials plot. The polynomials for different angles almost overlaps.

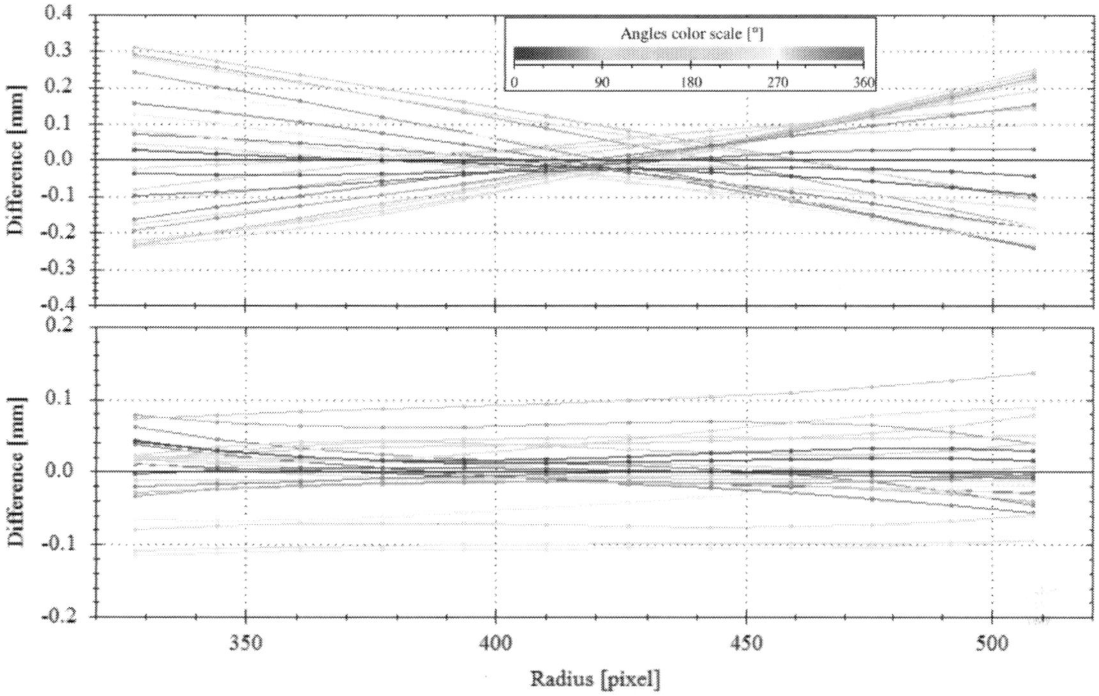


**Figure 10.** Calibration polynomials minus average polynomial plot for calibration aligning only at the 8th ring (top) and realigning for each ring (bottom). Plot of a polynomial for each 10°, total of 360.

*4.1.2 Measurement Error for Displaced SCG.* In order to evaluate the measurement error using the different calibrations (A and B), we captured images of the SCG displaced in the X and Y axis for different rings and positions. Figure 11 shows a photo of the setup and the acquired image of the 10th ring of the SCG displaced to the left. A circle was fitted to the measurement result in millimeters and the center of coordinate of the data displace to its center. The error was calculated subtracting the obtained moved radii from the respective calibrated radius of the SCG. The resultant errors using calibration 'A' and 'B', are shown, respectively, in figures 12 and 13. In figure 12, we also show the error using a mean polynomial for all angles for calibration 'A'. For calibration 'B' mean polynomial, the error plot was almost the same. The legends shows the fitted circle X and Y displacement in millimeters, the values were obtained using the mean polynomial.

Smaller errors were obtained using the calibration B. This result supports that the main factor of the differences of the polynomials was the misalignment of the trajectory of the SCG during calibration. In addition, the error using the mean polynomial for both calibrations were practically the same. This shows that the system has a small uneven behavior for different angles, except for regions around 220°, 275° and 330°. For each ring measured, the sum of the tendency plus two times the standard deviation was evaluated, the maximum value obtained is *Emax*. Table 1 shows an error comparison of the calibrations. In conclusion, a unique mean polynomial can be used in the cost of a small increase in the error. Also, an alternative solution would be to use an intermediate number of polynomials, since the polynomials don't vary significantly for close angles.

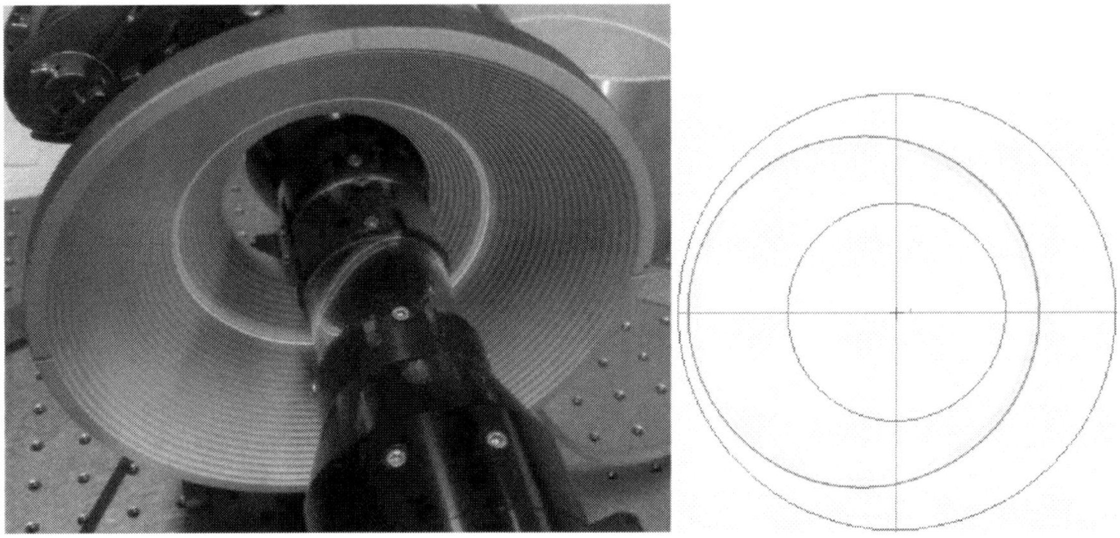

**Figure 11.** Photo of the profilometer measuring the 10$^{th}$ ring of the SCG displaced (left) and the resulting detected laser line (in red) at the measuring software (right). Concentric circles represent the radius measuring range.

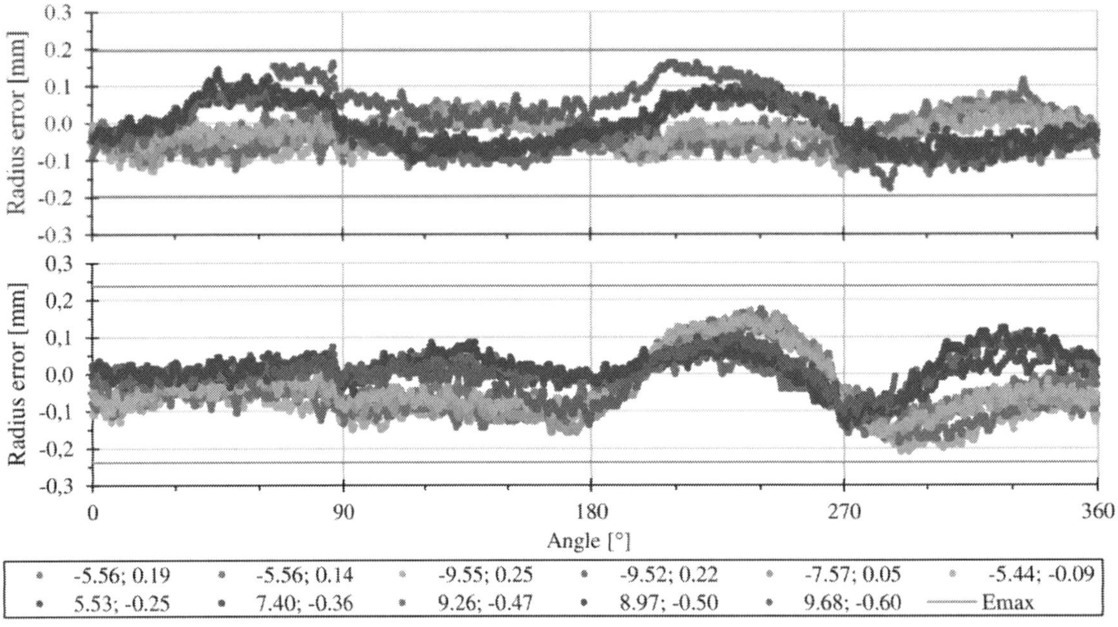


**Figure 12.** Error plot for different positions and rings. The calibration images were taken aligning the SCG for only ring 8$^{th}$. Polynomials fitted for each angle (top) and a unique mean polynomial for all angles (bottom).

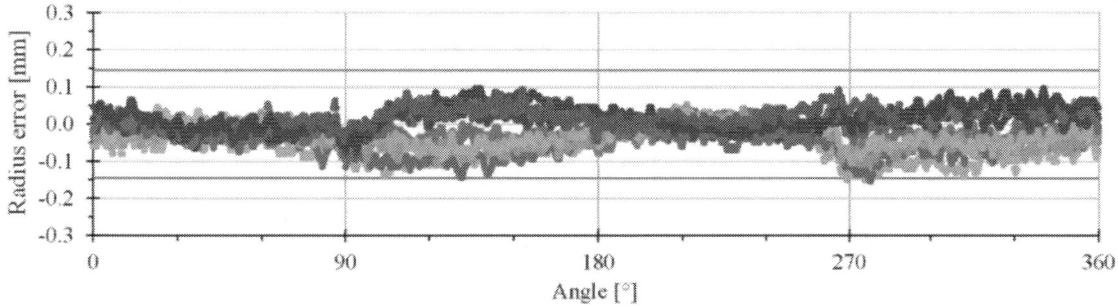


**Figure 13**. Error plot for different positions and rings. The calibration images were taken realigning the SCG for each ring. Polynomials fitted for each angle.

**Table 1.** Calibrations maximum error of the measurement of the SCG in various position comparison.

| Calibration | Procedure | *Emax* [mm] | |
|---|---|---|---|
| | | One polynomial for each angle | Unique mean polynomial |
| A | Alignment on the 8th ring | 0.20 | 0.24 |
| B | Re-alignment in all rings | 0.15 | 0.24 |

*4.2. Outside Laboratory Refinery Pipe Measurement*
Two refinery pipes were measured in field environment. Figure 14 shows a photo of the setup. Figure 15 shows the profilometer been inserted in one of the pipes. **Figure 16** shows two pictures taken by the profilometer, with the laser and LEDs on (left) and a typical measurement image (right).

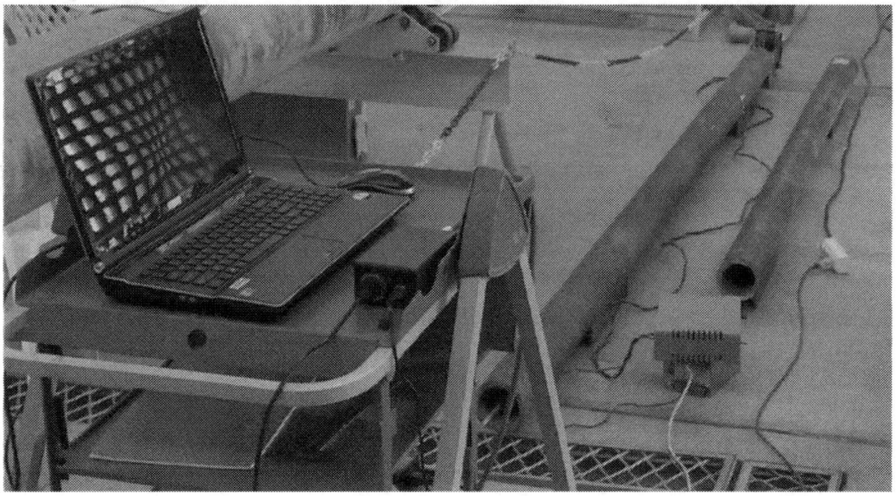

**Figure 14.** Refinery pipes measurement setup. Two 4” pipes can be seen on the right part of the photo.

**Figure 15.** Photos of the 4-inchprofilometer been inserted (left) and inside (right) the pipe.

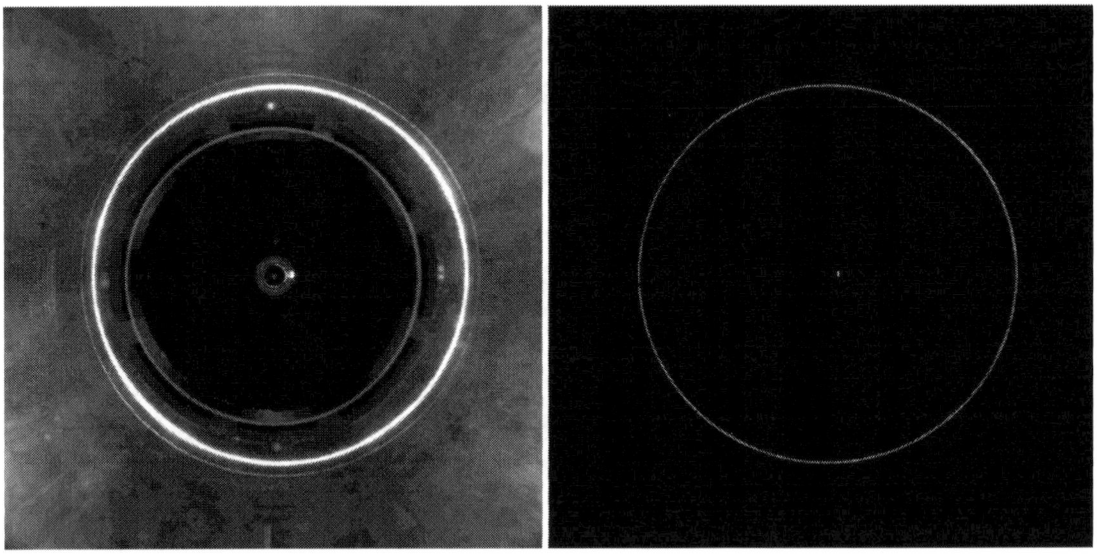

**Figure 16**. Photos taken by 4-inch profilometer inside a refinery pipe during the experiments. Laser and LEDs turned on, shutter speed set to 50 ms (left) and only laser turned on, shutter speed set to 2 ms (right).

The measurement results are displayed in three different forms: 1) as a colored 2D radii map, where the horizontal and the vertical axis represent, respectively, the Z axis, or the length, in millimeters and the angle in degrees; 2) as a graph of the mean diameter change in percent and 3) in 3D. The colors in the 2D and 3D images represent radii values.

The first pipe measured 'A' has a length of 2.2 m. The 2D analysis is shown in figure 17. The pipe did not present any localized corrosion or weld joint. Although it was detect a conical shape with very small diameter change of 0.35 % per meter.

Figure 18 shows the 2D analysis of the pipe 'B'. Also in this pipe, no corrosion was detected, but a pipe welded joint can be clearly seen around the middle of the graph. The inner diameter reduction in the welding is of about 1.5%. The 3D view of pipe 'A' and 'B' is shown in figure 19.

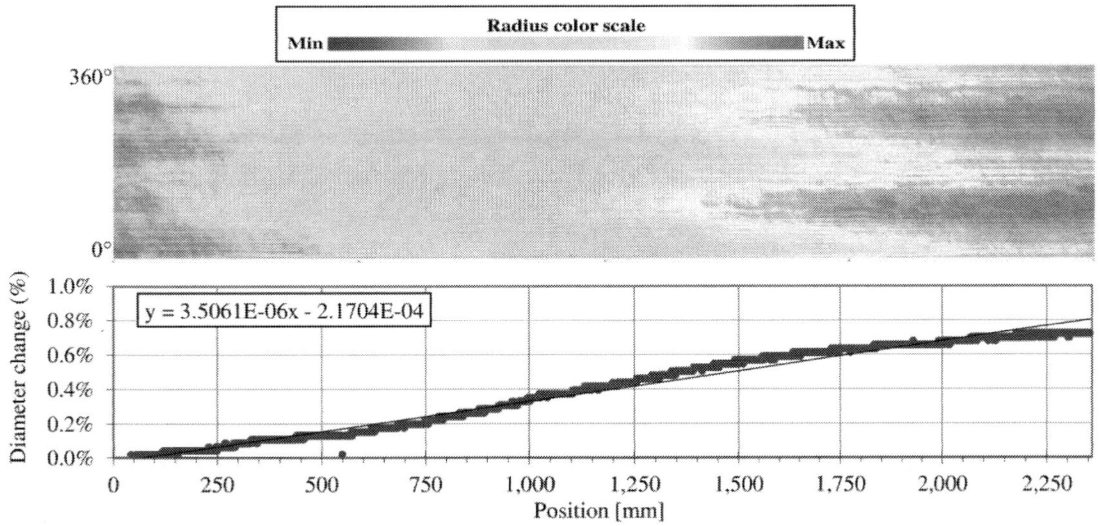


**Figure 17.** Pipe 'A' measurement result. Radius map and color scale (top); mean radius graph (bottom). Radius color range: 45.4 to 45.9 mm.

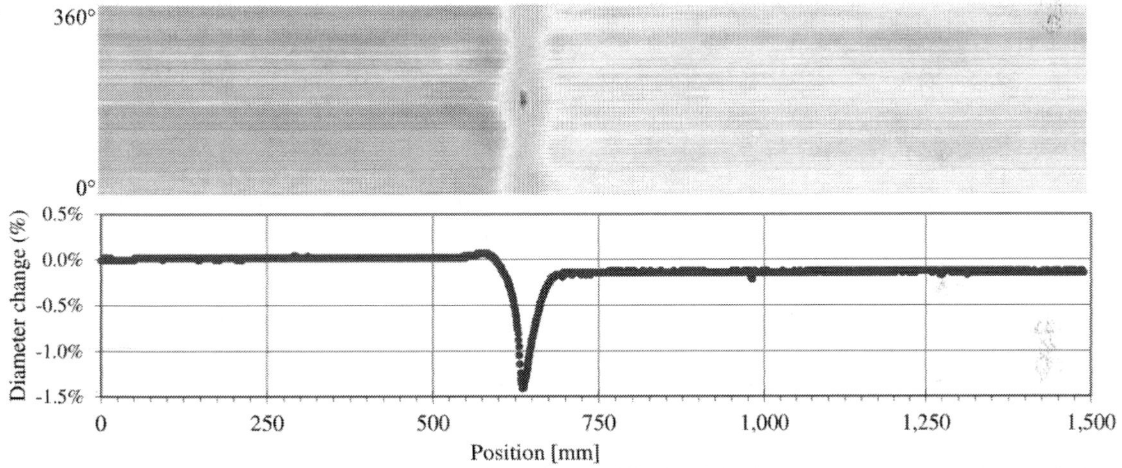


**Figure 18.** Pipe 'B' measurement result. Radius map (top) and mean radius graph (bottom). Pipe union around the middle of the measurement. Radius color range: 41.5 to 42.8 mm.

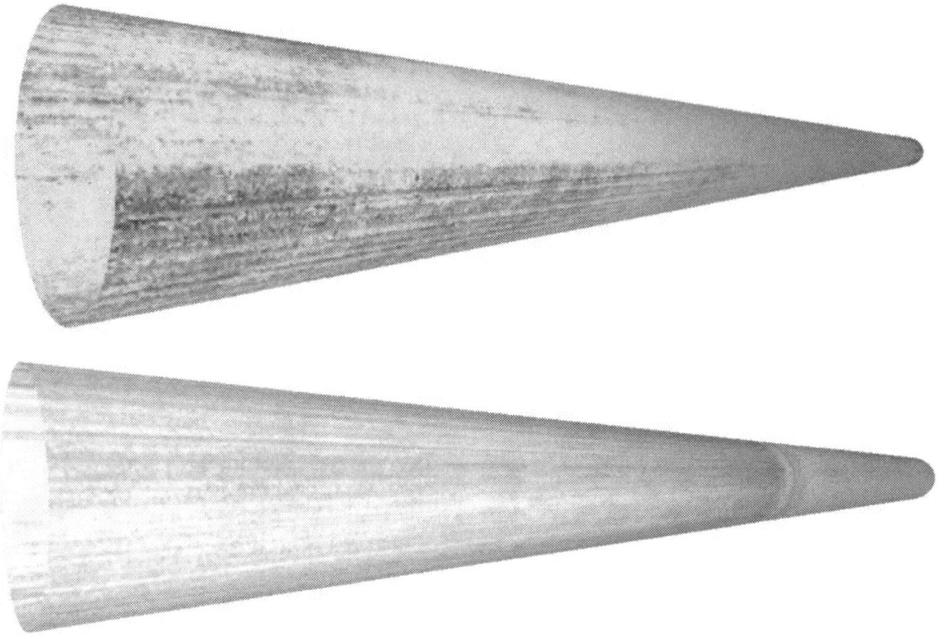

**Figure 19.** 3D view of the pipe 'A' (top) and 'B' (bottom).

## 5. Conclusion

We described a new measurement system for inner surface inspection of 100 mm (4") nominal diameter pipes that uses a radial laser light sheet emitter and a variation of the laser triangulation principle. Two calibrations procedures were tested and compared. A misalignment between the standard step cylindrical gauge and the profilometer was found to be the main factor of the differences in the polynomials for different angles. We evaluated that the radius measurement error in laboratory tests was of about ±0.2 mm.

The system was able to measure the inner surface of two refinery pipes with high resolution. On the first pipe, a variation in diameter of 0.35 % per meter was detected. On the second pipe, a welded union was measured, where the diameter is reduced by 1.5 %.

In future work, we will perform a deeper analysis of the profilometer error including a 3D measurement comparison. Also, different calibration approaches and the effects of the misalignment of the laser and the conical mirror must be investigated.

**Acknowledgments**

The authors would like to thank Petrobras/CENPES/TMEC for financing the optical laser profilometer project (PEROLA).

**References**

[1] Clarke T 1995 The development of an optical triangulation pipe profiling instrument *Opt. 3-D Meas. Tech. III-Applications Insp. Qual. Control Robot.* **1987** 331–40
[2] Hartrumpf M and Munser R 1997 Optical three-dimensional measurements by radially symmetric structured light projection *Appl. Opt.* **36** 2923
[3] Ritter M and Frey C W 2010 Rotating optical geometry sensor for inner pipe-surface reconstruction vol 7538, ed D Fofi and K S Niel pp 753803–753803 – 9
[4] Ravan M, Amineh R K, Koziel S, Nikolova N K and Reilly J P 2010 Sizing of 3-D Arbitrary Defects Using Magnetic Flux Leakage Measurements *IEEE Trans. Magn.* **46** 1024–33
[5] Davidson R 2002 An Introduction to Pipeline Pigging

[6] Camerini C, Weid J P von der, Freitas M and Salcedo T 2008 Feeler pig: a simple way to detect and size internal corrosion *IPC2008 7th International Pipeline Conference* (Calgary, Alberta, Canada) pp 1–9

[7] Inari T, Takashima K, Watanabe M and Fujimoto J 1994 Optical inspection system for the inner surface of a pipe using detection of circular images projected by a laser source *Measurement* **13** 99–106

[8] Zhuang B H and Zhang W 1998 Machine vision system for inner-wall surface inspection *Proc. SPIE 3306, Machine Vision Applications in Industrial Inspection VI* vol 3306, ed A R Rao and N S Chang pp 149–55

[9] Zhang G, He J and Li X 2005 3D vision inspection for internal surface based on circle structured light *Sensors Actuators A Phys.* **122** 68–75

[10] Wu E, Ke Y and Du B 2009 Noncontact laser inspection based on a PSD for the inner surface of minidiameter pipes *Instrum. Meas. IEEE Trans.* **58** 2169–73

[11] Yoshizawa T, Yamamoto M and Wakayama T 2006 Inner profile measurement of pipes and holes using a ring beam device *Proc. SPIE 6382, Two- and Three-Dimensional Methods for Inspection and Metrology IV* vol 6382, ed P S Huang p 63820D – 63820D – 6

[12] Wakayama T and Yoshizawa T 2008 Simultaneous measurement of internal and external profiles using a ring beam device *Proc. SPIE 7066, Two- and Three-Dimensional Methods for Inspection and Metrology VI* vol 7066, ed P S Huang, T Yoshizawa and K G Harding p 70660D – 70660D – 6

[13] Wakayama T, Machi K and Yoshizawa T 2012 Small size probe for inner profile measurement of pipes using optical fiber ring beam device *Proc. SPIE 8563, Optical Metrology and Inspection for Industrial Applications II* vol 8563, ed K G Harding, P S Huang and T Yoshizawa p 85630L

[14] Buschinelli P D V. 2007 *Development of an optical system with conical triangulation for inspection of inner profile of pipes* (Universidade Federal de Santa Catarina [in portuguese])

[15] Buschinelli P D V., Melo J R C, Albertazzi A, Santos J M C and Camerini C S 2013 Optical profilometer using laser based conical triangulation for inspection of inner geometry of corroded pipes in cylindrical coordinates *Proc. SPIE 8788, Optical Measurement Systems for Industrial Inspection VIII* vol 8788, ed P H Lehmann, W Osten and A Albertazzi (Munich, Germany) p 87881H

[16] Buschinelli P D V., Landeira G a. B, Kohler C, Albertazzi A and Teixeira da Silva G B 2012 A GPU based real-time image processing for an axis-symmetrical optical laser triangulation system *Optics and Photonics 2012: Hybrid SystemsHybrid Systems* vol 8498, ed A A S Awwal and K M Iftekharuddin p 849802

[17] Matheson I B C 1990 A critical comparison of least absolute deviation fitting (robust) and least squares fitting: The importance of error distributions *Comput. Chem.* **14** 49–57

Journal of Physics: Conference Series **648** (2015) 012011 doi:10.1088/1742-6596/648/1/012011

# New Approach for Non-Contact Measurement Using Vision Probe

**P B Costa[1,2], F O Baldner[1,2], J F S Gomes[2], W S Barros[2] and F R Leta[1]**

[1] Fluminense Federal University – UFF; Mechanical Engineering Department – Laboratory of Dimensional and Computational Metrology
Passo da Pátria Street, 156 – Niterói – Rio de Janeiro – Brazil
[2] National Institute of Metrology, Quality and Technology - Inmetro
Nossa Senhora das Graças Avenue, 50 – Xerém – Duque de Caxias – Rio de Janeiro – Brazil

E-mail: pedrobcostarj@gmail.com

**Abstract.** This paper has the purpose of presenting a study to determine the accuracy of a method for image measurement called Null Detector. In this new method, instead of determining the distance between two borders by counting pixels or by using subpixel approximations, the image itself is shifted and the border of the object to be measured is acquired in different points of the image. By using numerical adjustments for a specific point of the image, the real location for the border is found. The advantage of this method is that the pixel length does not need to be calibrated since it does not involve counting pixels.

## 1. Introduction

With Computer Vision, machines can make decisions by analyzing images. Its usage in any given process generally obeys the following steps: acquisition; segmentation; feature extraction; recognition; classification and decision-making [1].

Currently, Computer Vision systems are largely employed in control, monitoring and also in inspection. For control and monitoring systems, the information obtained from an image can be qualitative or quantitative, with reasonable variations. In such cases, the applications may have the purpose of evaluating a product quality in industrial environments or enhancing security systems.

In quantitative applications, measurements are extracted directly from the images, with national metrology institutes leading research into measurement by image in different areas. The measurements may vary in accuracy from tenth of a millimeter [2] to a couple of nanometers [3,4].

In high accuracy applications, the most used type of equipment are the coordinate measuring machines that use an optic probe, called simply optical CMM. They have been gaining ground in the industry due to the possibility of measuring several standards and parts which cannot be measured by contact, such as line scales, circular apertures, grids and mechanical components with small dimensions, such as gears, screws, among others [5].

This paper has the purpose of presenting a methodology for implementing a new method among several image measurement methods, called Null Detector. Measurements of graded line scales will be made using an optical CMM in order to verify the accuracy and applicability of this new methodology.

## 2. Measurement by Image

To measure an object from an image is to determine the distance between two borders of said object. This type of measurement may be accomplished in several different ways:

### *2.1. Manual Method*

1) The border of the object is aligned tangent to a fixed reference, a crosshair as shown in figure 1 [6].
2) Step number 1 is then repeated for the other border of the object whose measurement must be determined.

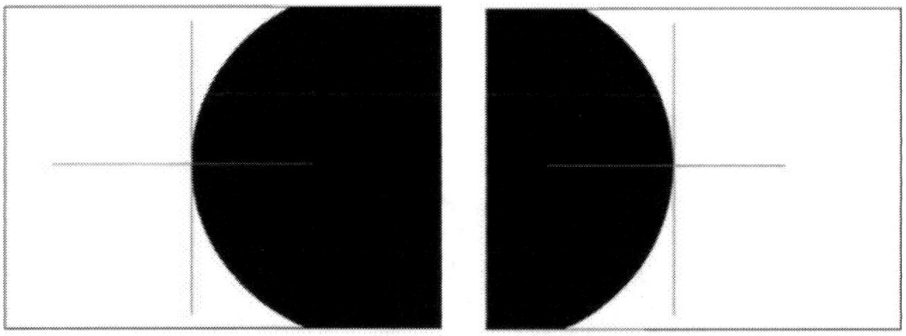

**Figure 1.** Diameter measurement using a fixed crosshair.

The measurement is then obtained by determining how much the object has been displaced.

### *2.2. Pixel Counting Method*

The main disadvantages of the manual method are the operator's experience in determining the object's border is directly tied to the accuracy and the error due to the crosshair thickness.

One way of eliminating the operator's influence is by using the pixel counting method [7]. In this method the measurement does not use a fixed reference in the image, but instead how much the object was displaced plus the difference between the two positions in the image, as can be seen in the example of figure 2.

1) From the first image the position for the border of the image and the displacement are acquired.
2) The same procedure of step number 1 is repeated in the object's other extremity.

The result of the measurement is given by adding the displacement of the object in the two images and the distance between the object's edges in both images. This distance has the pixel as a measurement unit, and, to obtain it in the proper length unit the pixel must be calibrated. This is achieved by using a dimensional standard with known uncertainties, which will, thus, add another contribution for the object's uncertainty of measurement.

## 3. Method

This section details the equipment and materials used for this experiment, as well as the methodology employed.

### *3.1. Measurement setup*

The Null Detector was used together with the measurement system for line scales of the Dimensional Metrology Laboratory (Lamed) of the Brazil's National Institute of Metrology, Quality and Technology (Inmetro).

The system comprises a coordinate measuring machine with an optical measurement system, as shown in figure 3.

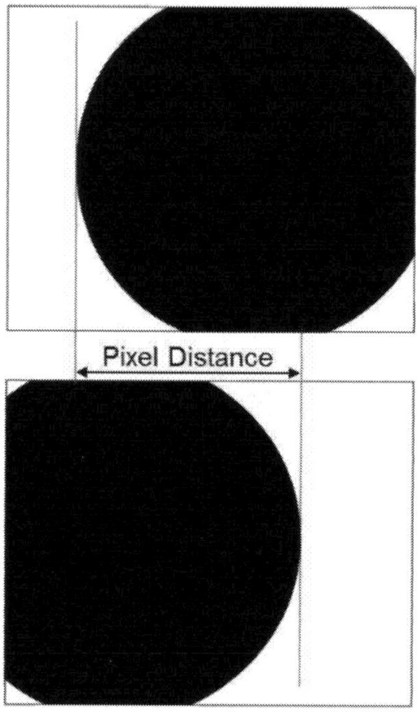


**Figure 2.** Diameter measurement by counting pixel.

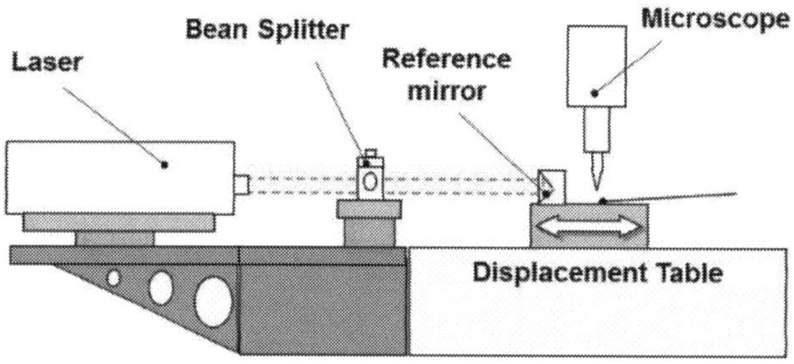


**Figure 3.** Line scale measurement setup.

In this system, the standard is place on the displacement table and aligned in relation to an interferometric laser. The displacements of the table are measured by the laser and are seen, at the same time, through the microscope with an magnification of up to 400x. These images seen through the microscope are acquired by a CCD camera.

*3.2. Null Detector*

The objective of the measurements with this new methodology is to eliminate the uncertainty contribution regarding the pixel calibration as means of decreasing the uncertainty of measurement as a whole for image measurement.

This methodology is an enhancement of the manual measurement previously described in item However, instead of having an operator manually position the extremities of the object on the crosshair show in the screen, these extremities are numerically projected to a fixed point.

For each of the object's extremities, a sequence of images is acquired, each in a different position. From every image the information for displacement and border position are extracted. With this set of data, two graphs can be plotted for pixel vs. displacement, one for each extremity. The sequence of points must be then fitted linearly, generating equations of the type $\boldsymbol{L} = \boldsymbol{a} \cdot \boldsymbol{p} + \boldsymbol{b}$ where position $L$, in millimeters, is to be determined for a fixed value of p, in pixel, that will be used on the two fitted equations that were calculated. The difference between the values of $L$ for the equations of both extremities is the measurement result.

### *3.3. Measurement Procedure*

A procedure has been designed to evaluate the Null Detector's performance by measuring a line scale of 5 mm with 1 mm as the smallest division.

For this procedure, the distances between the points of 1 mm, 2 mm, 3 mm, 4 mm e 5 mm will be measured with reference to the 0 mm point.

As stated in item 3.2, measurement with the Null Detector consists of acquiring a sequence of images. Five images were used for this test and, consequentially, five values for the displacement. When measuring the line scale, this procedure is repeated for each of the distances being measured. Thus, for every image both the pixel value for the center of the mark of the scale and the displacement value for the interferometric laser are extracted.

During measurement, five images are acquired and five displacement values are registered for the point of 0 mm. This is then repeated for all the other points, from 1 mm to 5 mm.

Equations for the displacement in function of the pixel value are interpolated for every measured point using the least squares method, obtaining:

$$\begin{cases} L_0 = a_0 + b_0 \cdot p \\ L_1 = a_1 + b_1 \cdot p \\ \quad \vdots \\ L_5 = a_5 + b_5 \cdot p \end{cases} \tag{1}$$

Where the indexes 0, 1, ..., 5 refer to the traces of the scale and the value of p is the point in the image chosen sit gat all the lengths are projected to it.

The graph below, in figure 4, shows the five curves obtained in one measurement cycle.

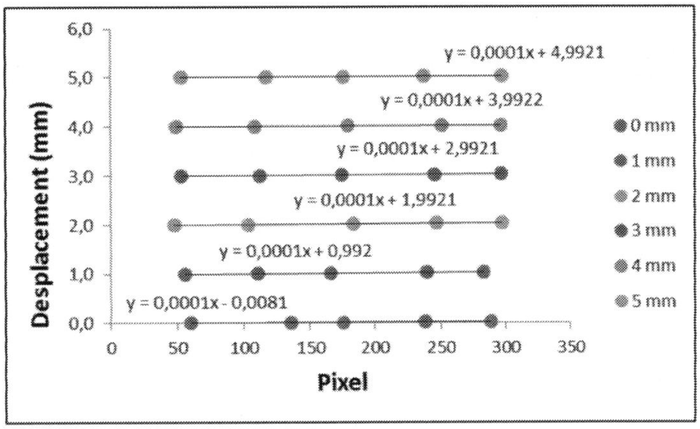


**Figure 4.** Graph of the points with the equations found for all the measured points.

Lastly, the lengths for the line scale are calculated by the difference between each calculated value of $L$ and the value of $L_0$, which is used as a reference.

## 4. Results

A line scale of 5 mm, with intervals of 1 mm, was measured in different days by using both the Null Detector Method and the traditional measurement procedure for line scales of the laboratory [8].

The graph in figure 5 shows the results of 5 measurements made in five different days using the Null Detector Method.

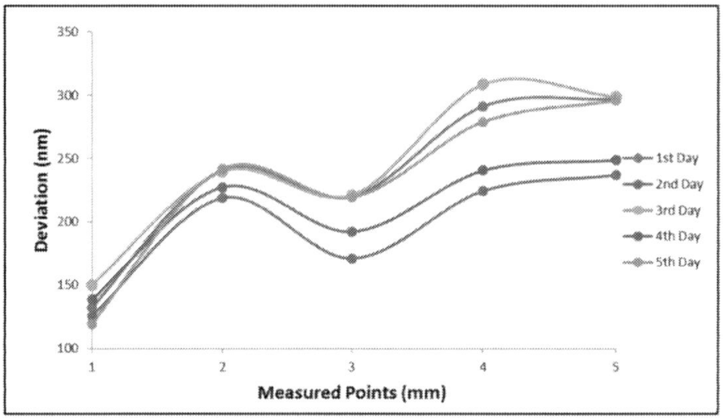


**Figure 5.** Deviation of the measured values.

The reproducibility of the system along the time can be observed in the graph of figure 5, where the biggest differences found are of approximately 60 nm in the 5 mm point.

Table 1 shows the comparison between the results obtained by the Null Detector and the laboratories standard procedure.

**Table 1.** Comparison of results between methods.

| Conventional Method | | Null Detector | | Normalized Error |
|---|---|---|---|---|
| **Deviation (nm)** | **U (nm)** | **Deviation (nm)** | **U (nm)** | |
| 113 | 80 | 132 | 50 | 0.20 |
| 236 | 80 | 241 | 50 | 0.05 |
| 209 | 80 | 220 | 50 | 0.11 |
| 250 | 80 | 291 | 50 | 0.44 |
| 284 | 80 | 297 | 50 | 0.14 |

A normalized error lower than 1 indicates compatibility between the results and it can also be seen that the uncertainties of measurement found are even lower than the ones from the conventional method [9].

In the conventional method, one of the most predominant sources of uncertainty is related to the length of the pixel, which, in this new method, is not needed. However, a new contribution for the uncertainty for this method is the one regarding the value of $p$, where all the lengths must be projected to. This uncertainty source is much reduced when the angular coefficients of the interpolated curves are low, thus decreasing the uncertainty of measurement as a whole.

The uncertainty component for the pixel length in the current is estimated to be of, approximately, 68 nm (for the tests shown in this paper), and the uncertainties of measurement found to be of around 80 nm. In the Null Detector method, the estimate influence of the parameter $p$ in the results is of around 5 nm, with the uncertainty of measurement being then reduced to 50 nm.

## 5. Conclusions

The analysis of the results for the measurement of the 5 mm line scale show good reproducibility of the system, with results being stable along different days. Looking further into the mathematical model for the calculation of the distances it is possible to conclude that the angular coefficient of the curves represents the pixel length in millimeter, and that the sensitivity of this method is directly related to this coefficient.

Besides the results shown in this article, experiments were also conducted for a magnification of 100x. For this magnification, however, the angular coefficient of the curves was approximately six times greater than the shown results. This leads to the conclusion that for lower magnifications the parameter p becomes a very large source of influence in the results.

So, it can be concluded that this method provides good accuracy with the shown uncertainties with devices that will provide small values for the pixel length and, consequentially, small angular coefficients.

**Acknowledgements**
The authors would like to acknowledge Faperj and Estácio de Sá University for their financial support under grants E-26/103.591/2012, E-26/103.618/2012 and E-26/171.362/2001, DPA – CI 461/2014

The authors would also like to acknowledge their colleagues from UFF and Inmetro for the support while conducting the experiments.

**References**

[1] Conci A, Azevedo E, Leta F R 2008 *Computação Gráfica: Teoria e Prática* (Rio de Janeiro: Editora Elsevier)

[2] Vieira G M, Leta F R, Costa P B, Braga S L, Gobin D 2010 Computer Vision Analysis of a Melting Interface Problem with Natural Convection *4th International Conference on Advanced Computational Engineering and Experimenting*

[3] Costa P B, Marques A, Baldner F O, Leta F R 2013 Line scale measurement using image registration *International Journal of Metrology and Quality Engineering*, **04**, 121-125

[4] Beers J S, Penzes W B 1999 The NIST Scale Interferometer *Journal of Research of the National Institute of Standards and Technology*, **104**

[5] Rodger G. 2001 Measurement Good Practice Guide: Dimensional Measurement Using Vision Systems *National Physical Laboratory*

[6] Costa P B, Barros W S, Oliveira R S, Dutra M S 2007 Development of a software for the line scales calibration system in INMETRO *The 19th Congress of Mechanical Engineering*

[7] Lassila A, Ikonen E, Riski K 1994 Interferometer for Calibration of Graduated Line Scales with a Moving Ccd Câmera as a Line Detector *Applied Optics*, **33**, 3600-3603

[8] Costa P B, Barros W S 2008 Calibração de Escalas Graduadas no Inmetro *1º Congresso Internacional de Metrologia Mecânica*

[9] Instituto Nacional de Metrologia, Normalização e Qualidade Industrial, Coordenação Geral de Acreditação 2011 Orientação Sobre Validação de Métodos Analíticos, *DOQ-CGCRE 008*

# Mathematical modeling of piezoresistive elements

**M Geremias**[1], **R C Moreira**[1], **L A Rasia**[1], **and A Moi**[1]
[1]DCEEng - Department of Physical Sciences and Engineering, *UNIJUI - Regional State University Northwest Rio Grande do Sul,*
3000 Street of Commerce, Ijui, RS, Brazil

E-mail: rasia@unijui.edu.br

**Abstract.** This article presents the longitudinal piezoresistive coefficients for thin film amorphous semiconductor type a-C:H. Experimental data and mathematical models have been used in computer simulations. The results show that a reduction of the longitudinal piezoresistive coefficient occurs due to the increased concentration of impurities in the films analyzed.

## 1. Introduction

The focus of this work are the piezoresistive sensing elements based on thin film semiconductor type diamond like carbon - DLC obtained by a system of magnetron reactive sputtering. They are used traditionally known equations for the monocrystalline and polycrystalline silicon to determine the value of the piezoresistive coefficients, $\pi_s$, amorphous DLC films and doped with nitrogen.

This paper presents the results of the piezoresistive effect for a film of hydrogenated amorphous carbon type (a-C:H/p-Si) with a nitrogen content ranging from 0% to 60%. These values are compared to models obtained by [1] into the monocrystalline silicon in which the impurity concentration values were obtained indirectly through mathematical simulations and experimental results given temperature range in the range 30 °C - 270 °C for film DLC [2].

## 2. Theory

The piezoresistive effect, $\frac{\Delta\rho_{ij}}{\rho}$, is the change in electrical resistance, R, of a semiconductor material when it is applied on a certain mechanical stress, $T_{xy}$ given by equation (1).

$$\frac{\Delta\rho_i}{\rho} = \sum_{j=1}^{6} \pi_{ij} T_j \tag{1}$$

The piezoresistive coefficients, $\pi_s$, for P-type and N-type material are related to the levels of doping impurity concentration, the crystallographic orientation of the material, temperature and the type conductivity [1,6] data for the matrix of equation (2).

[1] rasia@unijui.edu.br.

$$\pi_{ij} = \begin{pmatrix} \pi_{11} & \pi_{12} & \pi_{12} & 0 & 0 & 0 \\ \pi_{12} & \pi_{11} & \pi_{12} & 0 & 0 & 0 \\ \pi_{12} & \pi_{12} & \pi_{11} & 0 & 0 & 0 \\ 0 & 0 & 0 & \pi_{44} & 0 & 0 \\ 0 & 0 & 0 & 0 & \pi_{44} & 0 \\ 0 & 0 & 0 & 0 & 0 & \pi_{44} \end{pmatrix} \quad (2)$$

In many cases, the mechanical stress, $T_{ij}$, are components along the axis of the semiconductor crystal, but in other cases it works on an arbitrarily oriented coordinate system, so that the generalized Hooke's law should be used to determine the elastic deformation coefficients, $S_{ijkl}$, expressed by equation (3).

$$\varepsilon_{ij} = S_{ijkl}.T_{ij} \quad (3)$$

Since, $S_{ijkl}$, is a fourth order tensor constant elastic deformations of the substrate or film production considered for the sensor elements or diaphragms.

In matrix form, Equation (3) can be written according to equation (4).

$$\begin{pmatrix} \varepsilon_1 \\ \varepsilon_2 \\ \varepsilon_3 \\ \varepsilon_4 \\ \varepsilon_5 \\ \varepsilon_6 \end{pmatrix} = \begin{pmatrix} S_{11} & S_{12} & S_{12} & 0 & 0 & 0 \\ S_{12} & S_{11} & S_{12} & 0 & 0 & 0 \\ S_{12} & S_{12} & S_{11} & 0 & 0 & 0 \\ 0 & 0 & 0 & S_{44} & 0 & 0 \\ 0 & 0 & 0 & 0 & S_{44} & 0 \\ 0 & 0 & 0 & 0 & 0 & S_{44} \end{pmatrix} \begin{pmatrix} T_1 \\ T_2 \\ T_3 \\ T_4 \\ T_5 \\ T_6 \end{pmatrix} \quad (4)$$

The piezoresistive coefficients and the mechanical stress can be converted to mechanical deformation by using the Young's modulus, $E$, [3,4] according to equation (5).

$$E = \frac{T}{\varepsilon} \quad (5)$$

Where, $\varepsilon$, is the longitudinal elastic deformation of the test body (dimensionless).

In technical terms, numerically, load and pressure sensors can be described by equation (6).

$$GF = \frac{\Delta R}{R\varepsilon} \quad (6)$$

The literature provides information on many of the properties of materials, however, it is up to the designer to evaluate and determine those that are important for a specific project. In this work were carried out reviews of relevant characteristics for a project of sensor elements for piezoresistive pressure transducers.

Mechanical sensitivity is changed mainly by the material resistivity, $\rho$, mobility, $\mu$, temperature, $\theta$, and concentration of impurities, N. The piezoresistance coefficients dependent on temperature and concentration of dopants. Therefore, the electrical properties of piezoresistores sensors are extremely sensitive to the presence of impurities, even in small concentrations [2]. Since the impurity concentration is given by equation (7).

$$N = \frac{\rho}{q\mu} \quad (7)$$

The diagram shown in figure 1 shows how a semiconductor material behaves in electrical terms, mechanical and thermal [2]. From the analysis of these properties begins the design of a transducer.

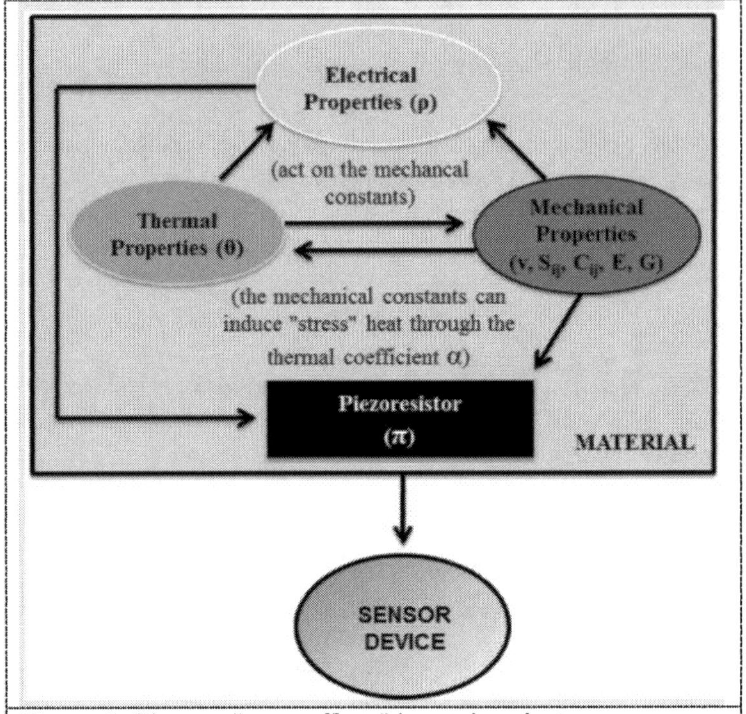


**Figure 1.** Diagram "cross effects" in semiconductors.

In addition to the dependencies represented in the diagram of figure 1, the sign and magnitude of the piezoresistive effect in semiconductor materials are associated with the dopant type, temperature, and quality crystallographic orientation [6].

## 3. Features DLC films

The diamond-like carbon films are amorphous materials with a structure in which carbon atoms bonds vary between graphite and diamond. These materials have unusual combination of electrical, optical, thermal and mechanical [7]. Its main properties are: hardness, high carrier density, low roughness, chemical and inherent possibility of being intrinsically produced, p-type and n-type with a band gap ranging from 0.5 eV to near 4 eV [8].

In addition to presenting high strength, the films a-C:H show good step coverage can be used as anti-corrosion barriers and have the advantage of being filed with low roughness and excellent uniformity over large areas [10]. But have disadvantages such as high voltage and low internal mechanical stability which can be minimized considerably through improvements in deposition techniques. These strongly influence the electrical characteristics of DLC films. Since only changing the temperature of deposition can vary the resistivity $10^6$ to $10^{16}$ $\Omega.cm$ [8].

The piezoresistive effect on DLC films can be defined as an agglomerated composition $sp^2$ erratically in an insulating matrix $sp^3$ as shown in figure 2. In this description, the agglomerates $\pi$ bonding energy with no local create conditions for $sp^2$ bonds form the first pair, and subsequently forming C-C bonds (ethylene). The recombinations of bonds then form hexagonal rings such as benzene. Following the masses become graphitic within the $sp^3$ matrix.

In figure 2, the size of the agglomerates is indicated by, $s$, which are relatively large and the distance between two clusters is indicated by, $d$, which is typical of quantum mechanical tunnelling.

Similarly, the model for many valleys silicon, where energy ellipsoids are deformed when a mechanical stress is applied this model to the a-C:H is observed that when a mechanical stress is applied to the piezoresistor, the value of distance between two clusters $sp^2$ increases and therefore the amount of piezoresistance also increases. Otherwise, when a compressive mechanical stress the distance between the clusters is being applied to reduce the value of this decrease is attached piezoresistance [2].

The piezoresistive effect described in this way follows the model that explains the sensitivity factor or "gauge factor" (GF) in thick film resistors [7], establishing the empirical relationship between the value of electrical resistance and sensitivity. The very resistive materials should have high GF values. Otherwise, the source of the piezoresistive effect on DLC films can be described as a composition $sp^2$ agglomerates unevenly in a $sp^3$ matrix insulator as shown in figure 2.

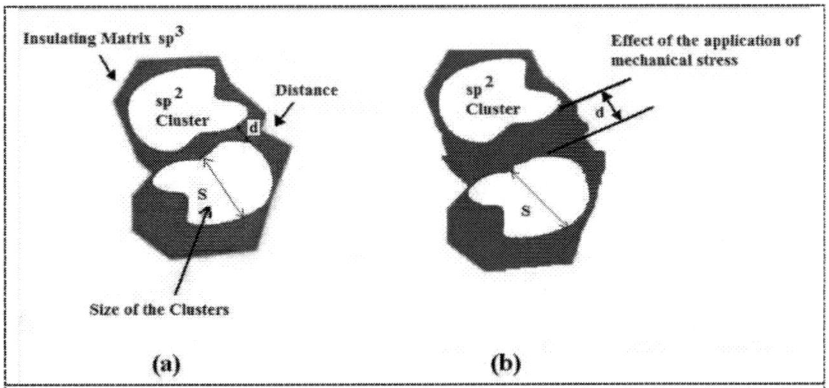


**Figure 2.** Model for DLC films. (a) without mechanical and (b) with mechanical stress effort [2].

As there are many similarities between the various DLC films used as piezoresistive elements can approach them to the models equivalent to those of silicon with cubic structure and shape the practical way of sensing element [2].

## 4. Mathematical model

It is noted that the sign and the magnitude of the piezoresistive effect in thin films associated with semiconductor dopant type, temperature, crystallographic orientation and quality [2,6]. Therefore, all of the physical properties of a material are dependent on the temperature so that the physical designs of piezoresistores must take this fact into consideration as shown in equation (8).

$$R(P,\theta) = R_{ref}(\theta)\left[1 + \pi_{ll}(\theta)T_{ll}(P,\theta) + + \pi_t(\theta)T_t(P,\theta) + \pi_{xy}(\theta)T_{xy}(P,\theta)\right] \quad (8)$$

Where, $P$ is the applied pressure, $\theta$ is the temperature, $R_{ref}$ is no mechanical stress resistance at the reference temperature (300K), $\pi_{xy}$ piezoresistive coefficient is the shear in the xy plane and $T_{xy}$ is the average shear mechanical stress [2] .

## 5. Results and discussions

Using the experimental data obtained by [2] for a film type a-C:H/Si/Ag with average resistivity, $\rho_{avg}$ = 4.22 MΩ.cm was performed mathematical simulation to obtain the longitudinal piezoresistive coefficients for the film. The values were compared with films containing a nitrogen level which increase the field emission properties and reduce internal mechanical stress in the films.

The DLC films are characterized by having high resistivity values in a very wide range. Therefore, if they were designed with piezoresistor 2500 Ω typical resistance as is usual with mono and polycrystalline silicon films, the films analyzed in current research should have an extraordinary level of doping as illustrated by the results shown in the next figures.

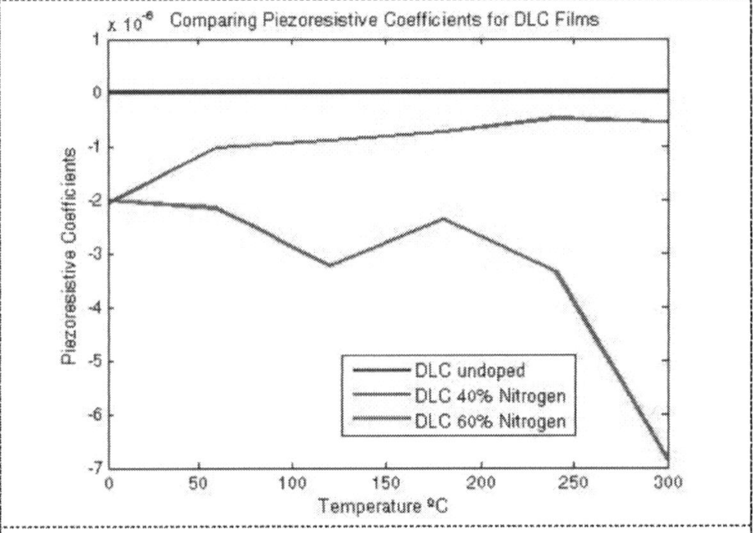


**Figure 3.** Longitudinal piezoresistive coefficient for different DLC films (doped and undoped).

The variation of the longitudinal piezoresistive coefficient is best observed at temperatures above 100 °C the films doped with 40 % and 60 % of nitrogen. In certain temperature ranges these doped layers have increasing linear behaviour and other decreasing this occurs because of changes that the dopant and the temperature produce the internal properties of the film. Therefore, the linear behaviour shown in Figure 3, the undoped DLC film, is according to the values suggested by the literature.

Figure 4 shows the function of the resistance value variation for doping the film a-C:H undoped (that is without any content of nitrogen) and temperature increases in the range of approximately 45 °C.

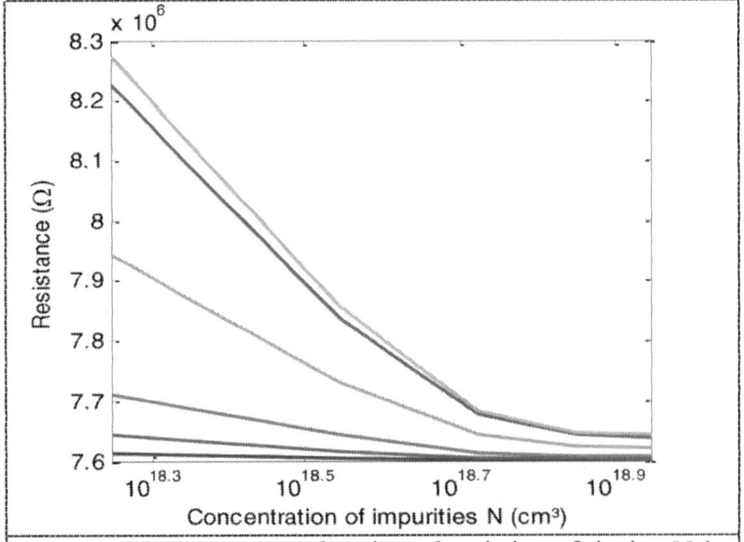


**Figure 4.** Resistance as a function of variation of doping N in a-C:H undoped (label to the film without nitrogen content).

Figure 4 shows the exponential function of the behaviour of the resistance variation in doping undoped DLC films, this behaviour is in agreement with what is suggested in the literature for undoped Si films compared with films analyzed in this study. There is also in figure 4 that depending on the manufacturing processes the piezoresistor tends to assume a constant average value of electrical resistance as suggested [2].

Figure 5 shows the resistance value of the film a-C:H doped with 40 % nitrogen and temperature increases in the range of 33 °C.

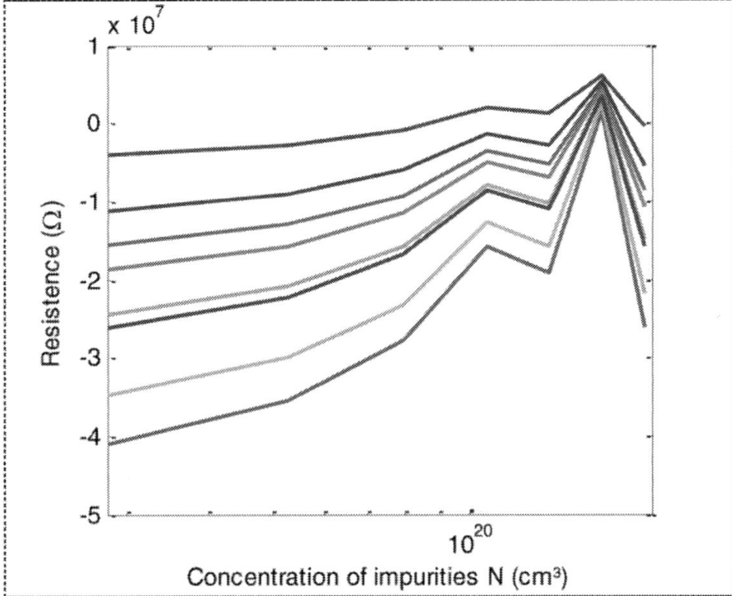


**Figure 5.** Resistance as a function of variation of doping N in a-C:H doped with 40 % of nitrogen.

The electrical behaviour of the film, shown in figure 5, is due to the changes caused by the incorporation of nitrogen. The appearance of the curve also varies with the temperature and compared to the results shown in figure 4, show a significant increase.

Figure 6 shows the resistance value in function of the variation of doping to the film a-C:H with 60 % nitrogen and which increases the temperature in the approximate range of 45 °C.

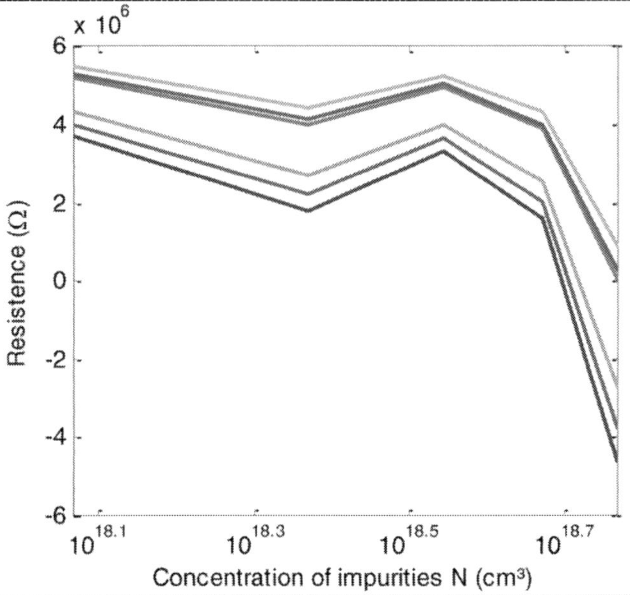


**Figure 6.** Resistance as a function of variation of doping N in a-C: -H doped with 60 % of nitrogen.

Comparing the values shown in figure 6 with Figure 4 clearly noticed an increase of the resistivity in the first figure. Now, when comparing these values with those shown in figure 5 it is evident the influence of the doping concentration in the resistance value of the material changes dramatically including the shape of the curve.

Because of this analysis it is up to the designer piezoresistors evaluate and determine the amount of doping ideas for a specific project.

The change in electrical resistance for the film a-C:H undoped, subject to mechanical stress parallel to the flow of electric current is called the longitudinal piezoresistive coefficient, $\pi_l$, and is being shown in figure 7.

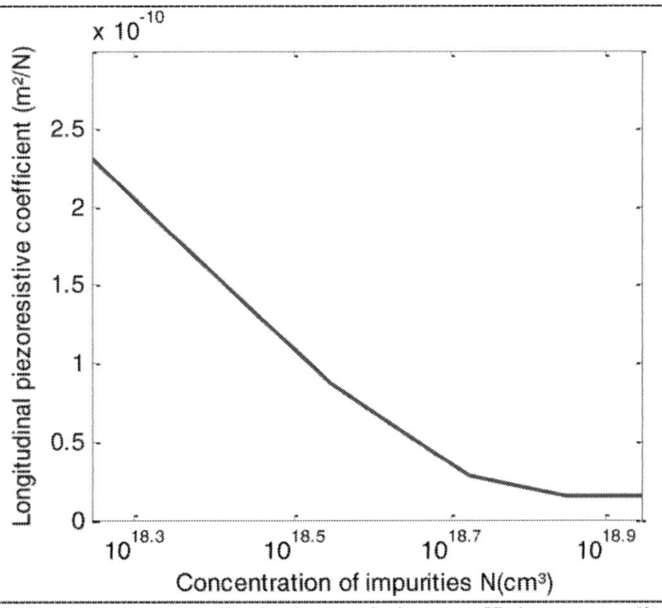


**Figure 7.** Longitudinal piezoresistive coefficient according to the N doping of variation for a-C:H undoped (label to the film without nitrogen content).

The exponential behaviour shown in figure 7 for the longitudinal piezoresistive coefficient versus doping variation is characteristic for type semiconductor film undoped a-C:H. The longitudinal piezoresistive coefficient shown in figure 7 tends to assume a constant average value as it gradually increases the value of the doping concentration.

The behaviour of the piezoresistive coefficient, $\pi_l$ , shown in figure 8 is the film a-C:H doped with 40 % of nitrogen.

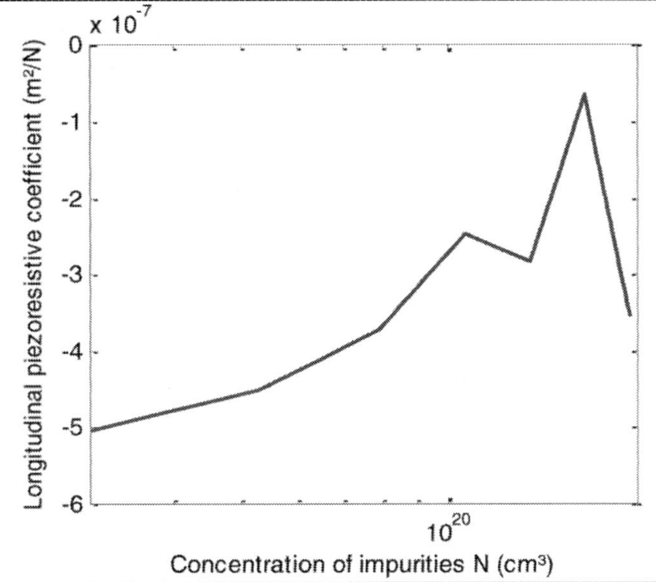


**Figure 8.** Longitudinal piezoresistive coefficient according to the N doping of variation for a-C:H doped with 40 % of nitrogen.

Figure 8 shows a considerable increase in the value of the longitudinal piezoresistive coefficient in relation to figure 7 when comparing values of incorporating the dopant in the film.

Figure 9 shows the value of the piezoresistive coefficient, $\pi_l$, to the film a-C:H doped with 60 % of nitrogen.

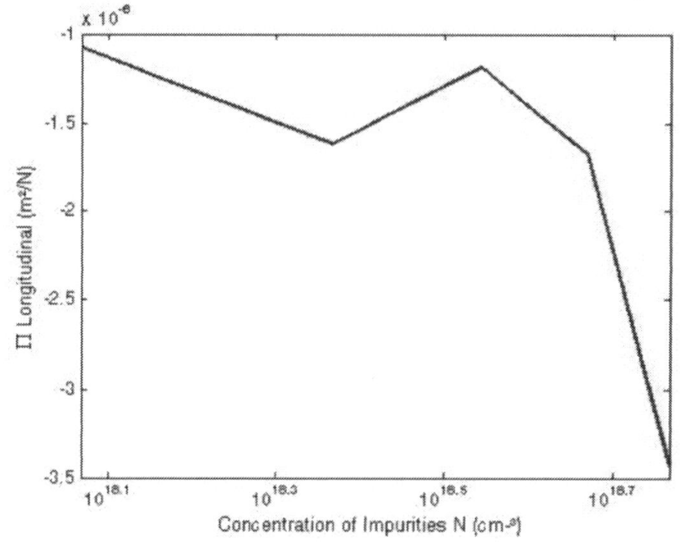


**Figure 9.** Longitudinal piezoresistive coefficient according to the N doping of variation for a-C:H doped with 60 % of nitrogen.

It can be seen in figure 9 that the longitudinal piezoresistive coefficient has higher values when compared to figure 7 and figure 8. Therefore, it is evident that the nitrogen concentration significantly influences the piezoresistive coefficient of the material.

The results shown indicate that can be estimated with good precision the behaviour of a piezoresistive sensor element to be designed, although the values of the coefficients are relatively small.

## 6. Conclusion

This work shows that are needed new simulations to improve the models set to obtain the piezoresistive coefficients for thin film type doped DLC. It was observed that there is some coherence between the results obtained with those proposed in the literature. In many situations, particularly those in which the film is doped models proposed for crystalline films can be applied.

## 7. References

[1] Kanda Y, 1982 A graphical representation of the piezoresistance coefficients in silicon, *IEEE Transactions on Electron Devices*, ED-29 **1** pp 64 - 70

[2] Rasia L A, 2009 Study and application of electrical, thermal and mechanical properties of amorphous materials in piezoresistive pressure transducers São Paulo (Doctoral Thesis), p 190 Polytechnic School of the University of São Paulo, São Paulo, Brazil

[3] Richter J, Hansen O, Larsen A N, Hansen J L, Eriksen G F, Thomsen E V 2005 Piezoresistance of silicon and strained Si0.9Ge0.1, *Sensors and Actuators*, A **123-124 C** pp 388 – 396

[4] Dally, J W, Riley W F, 1978 *Experimental stress analysis*. McGraw-Hill Inc Second Edition pp 205-214

[5] Luthje H, Biehl S, Bandorf R, Sick J H, Peiner E, Tibrewala A, 2005 Preparation and characterization of multifunctional thin film sensors based on amorphous diamond-like-carbon. *The 13th Int. Conf. on Solid-State Sensors, Actuators and Microsystems*, Seoul, Korea

[6] Maisel, Li e Glang, R, *Handbook of thin film technology*. McGraw-Hillbook Company, 1983.

[7] Grill, A, 1999 Diamond-like carbon: state of the art. *Diamond and Related Materials*. **8** pp 428-434

[8] Santos, A M, 2005 Development and application processes with high density plasma deposition of carbon films (Master's thesis) University of São Paulo, São Paulo, Brazil

[9] Kelires, P C, 2001 Stress properties of diamond-like amorphous carbon. *Physics* B, **296** pp 156-162

Journal of Physics: Conference Series **648** (2015) 012013 doi:10.1088/1742-6596/648/1/012013

# An imaging grating diffractometer for traceable calibration of grating pitch in the range 20 µm to 350 nm

**D A Brasil[1], J A P Alves[1], J R Pekelsky[2]**

[1]Dimensional Metrology Laboratory (Lamed), Mechanical Metrology Division (Dimec), Directory of Scientific and Industrial Metrology (Dimci), National Institute of Metrology, Quality and Technology (Inmetro), Av. Nossa Senhora das Graças, 50, Xerém, Duque de Caxias, RJ, Brazil, ZIP 25250-020
[2] Canadian Dimensional Metrology, Ottawa, Canada

E-mail: dabrasil@inmetro.gov.br

**Abstract**. This work describes the development of a grating diffratometer to provide traceable calibration of grating pitch in range 20 µm to 350 nm. The approach is based on the Littrow configuration in which a laser beam is directed onto the grating which is mounted on a rotary table and can be turned so that each selected diffraction order is retro-reflected in the laser incidence direction. A beamsplitter and a lens direct the reflected diffraction order to form a small image spot on a CCD camera and the spot centering is used to adjust to rotation angle, thereby giving the diffraction angle. Knowing the diffraction angle for several orders and the wavelength of the laser, the average grating pitch can be determined to an uncertainty the order of 14 pm.

## 1. Introduction

EURAMET is an important Regional Metrology Organization (RMO), which forecasts technology market demands and makes roadmaps of quantities. Its length metrology committee is always looking for ways of develop methodologies capable to deal with measurements of dimensions that get smaller and smaller each day [1]. This is because industrial processes need tighter tolerances to improve quality and performance. As a consequence, eventually, National Metrology Institutes (NMIs) face paradigm breaks whenever new technologies emerge, so they need to implement different methodologies to measure with higher precision.

Optical instruments have experienced a huge development in the last decades with the parallel improvements in microelectronics and automation as well as the incorporation of digital image processing [2-3]. No matter how technological is the instrument, it needs to be calibrated to keep up with the international standardization. Regarding optical instruments, in general they need special gages on which there are engraved lines in one or two dimensions, regularly spaced in dimensions in the micrometric order. They are called diffraction gratings and are shown in figure 1 and are characterized by its average pitch which is calculated over the full area of the surface of the grating. For the calibration of pitch of this order of magnitude, it can be done with a relatively simple and inexpensive system that operates with the optical phenomenon of diffraction. This system allows calibration of gratings with distances between lines in the order of the magnitude of the wavelength of light used to produce the phenomenon, which is, in case of common He-Ne, about 0.6 µm.

(a) (b)

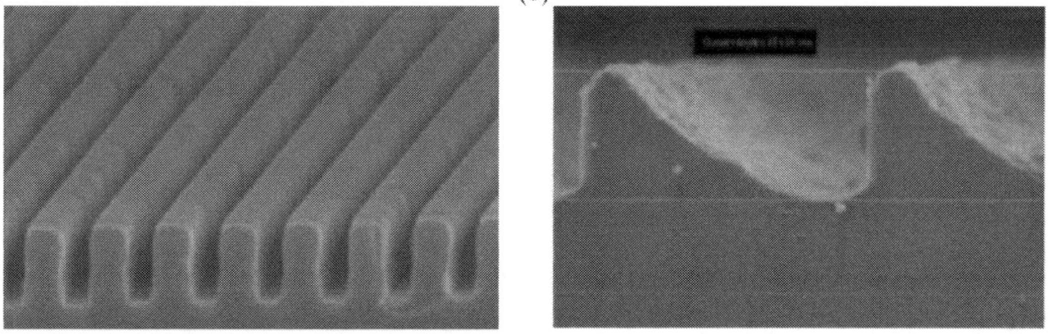

**Figure 1.** Scanning electron microscope (SEM) image of (a) square profile and (b) a ramped grating structure. Both are suitable for pitch calibration by diffractometry [4]

Besides used as standards for optical microscopy, the gratings may be used as reference to calibrate atomic force microscopes (AFM) concerning its horizontal scale. An AFM is an instrument capable of capturing the profile of a surface by scanning a very small tip built in the end of a cantilever whose movement is detected by laser. When the system moves in a direction parallel to the surface, the tip is submitted to deflections due to forces in atomic magnitude so that profiles are obtained with a very high resolution. Those equipment can be found in the market (figure 2.a) and to calibrate them is mandatory before putting them into operation. There are basically two fundamental measurands that must be calibrated: the horizontal distance and the vertical displacement of the tip. Some gages can be used for this purpose, among them, gratings, but calibrating the grating remains a challenge.

NMIs have designed and mounted AFMs to enable its traceability in both vertical and horizontal movement and these are called metrological AFMs (figure 2.b). The task of building an AFM that is intrinsically traceable involves a huge amount of time and resources. Other alternative is to adapt a conventional AFM in order to make it metrological. But even in this case, the complexity is considerable. Both types can be used to calibrate the gratings and disseminate the traceability to other lower quality instruments.

(a) (b)

**Figure 2.** Types of AFM: (a) commercial [5] and (b) metrological [6]

Journal of Physics: Conference Series **648** (2015) 012013 doi:10.1088/1742-6596/648/1/012013

There is, however, a quite big difference between the calibration by using an AFM and the principle of diffratometry: in the first, the pitch P is evaluated at a very localized spot and the path covered by the tip of sensor is in the order of micrometers. In the second, a much larger surface is covered by the spot of laser light so that an average value for the pitch is obtained using a very large number of grating lines.

The idea behind the diffractometer method is that the incident light in a grating will be diffracted back by retroreflection just in some specific angular position of the grating. First, the grating is positioned so that the reflection signal for the zero order is obtained. Then, the grating is rotated relatively to the axis of the laser beam by a sufficiently precise angular position, using for instance a precise rotary table, until a new point of maximum reflection intensity appears. The bright points can be imaged into a CCD camera which works as a null detector. So, from a bright spot correspondent to the zero order to another one correspondent to the $m^{th}$ diffraction order, an angle is realized and can be associated to the pitch of the grating. To each position, is associated a mode of diffraction and in those positions the incident and reflected light are coaxial (the Littrow configuration).

NMIs in various countries have been developing and assembling diffractometers, such as NRC in Canada, PTB in Germany, METAS in Switzerland and CENAM in Mexico, to name a few. They basically differ in the detection of the beam: METAS uses a quadrant detector, CENAM and NRC a CCD camera, and PTB a mix of both.

At METAS measurements are possible to step from 20 μm down to 320 nm for red light He-Ne laser (633 nm) and below 275 nm for green light laser He-Ne (543 nm). At PTB they use three sources of lasers with a wavelength range of 633 nm, 543 nm and 266 nm (UV), allowing reticule calibration in the range 150 nm to 10 μm. Depending on the pattern quality, PTB can achieve an uncertainty smaller than 10 pm, and smaller than 6 pm can be obtained at METAS [7-9]. NRC developed a methodology which resembles that developed by METAS, as mentioned above. According to study reported NRC, three different grating pitches of 700 nm, 4 μm and 10 μm were measured using a laser source wavelength of 633 nm, to uncertainties of about 7 pm [10].

The use of diffraction phenomenon to calibrate the distances between grating lines is the path taken by many NMIs because of its simple methodology and low cost of implementation and operation compared with a metrological AFM. It fits InmetroO´s goals since it opens room for the development of methods for fine-pitch grating calibration by using light phenomena associated with high accuracy at low cost.

## 2. Methodology

The schematic drawing of the proposed diffractometer is presented in figure 3 and the CAD drawing in figure 4 and was modeled after NRC imaging diffractometer design. The main concept is to use the Littrow configuration in grating mounting and observe each angle θ relative to the zero order angle of the rotary table, figure 5, that reach a reproducible common position of the diffracted spot in a CCD camera (null detector mode) [10]. With these results and the wavelength of the monochromatic laser λ, the average pitch P can be determined for the $m^{th}$ order by the equation:

$$P = \frac{m^{th}\lambda}{2\sin\theta} \tag{1}$$

The pitch found is an average result because the laser beam covers an area spanning many pitches of the grating. The retro-diffracted beam creates a pattern image for diffraction order composed by each retro-diffracted pitch of the same order.

The Littrow configuration is a special grating mounting in which the grating is positioned with incident beam coincident with the retro-diffracted spot. The zero order, figure 5(a), is the base position whose return beam has higher intensity. To reach Littrow configuration of other orders, the diffraction grating is mounted on a precision rotary table and properly aligned. In the figure 5(b) the first order was obtained turning the rotary table to reflect the diffracted beam back along the direction of the incident beam.

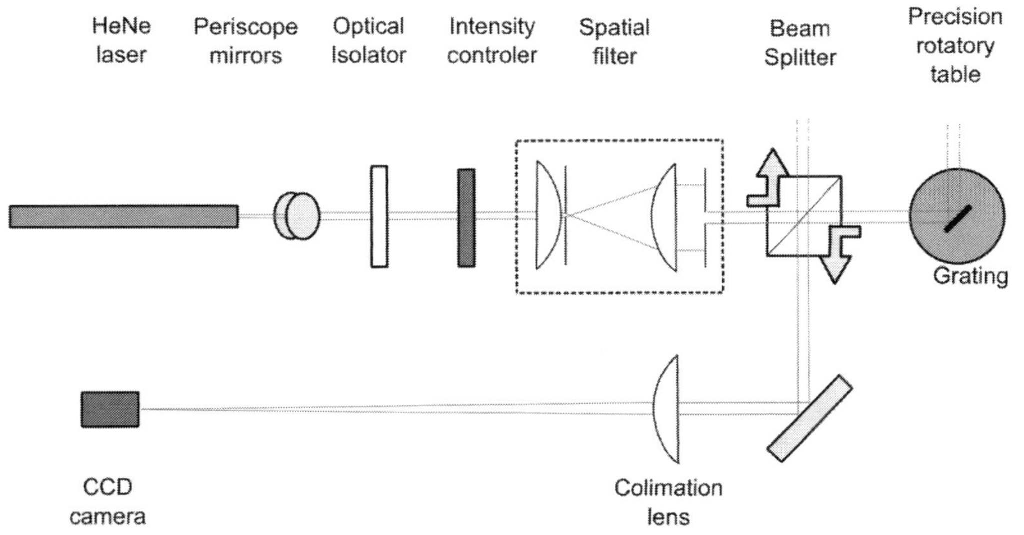


**Figure 3.** Inmetro diffractometer schematic drawing

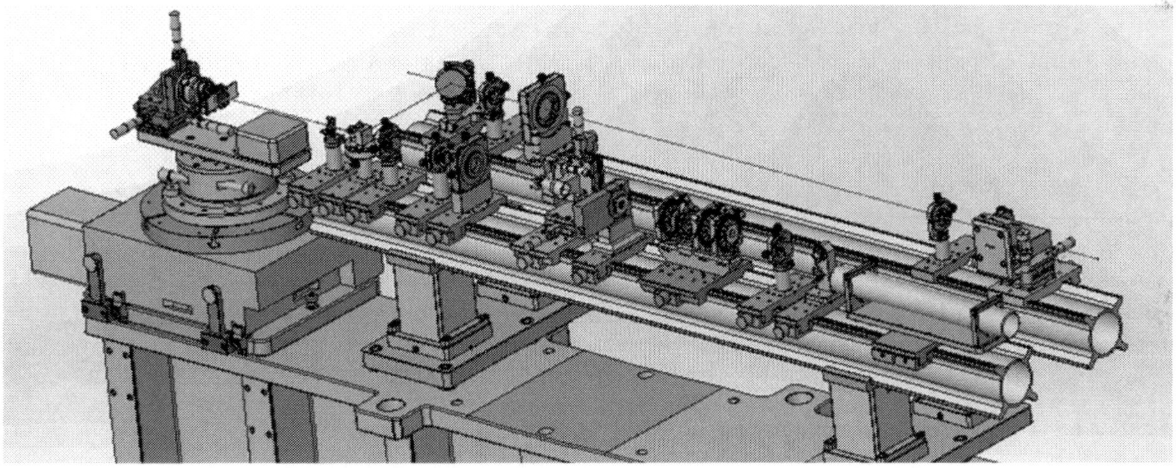

**Figure 4.** Inmetro diffractometer CAD drawing

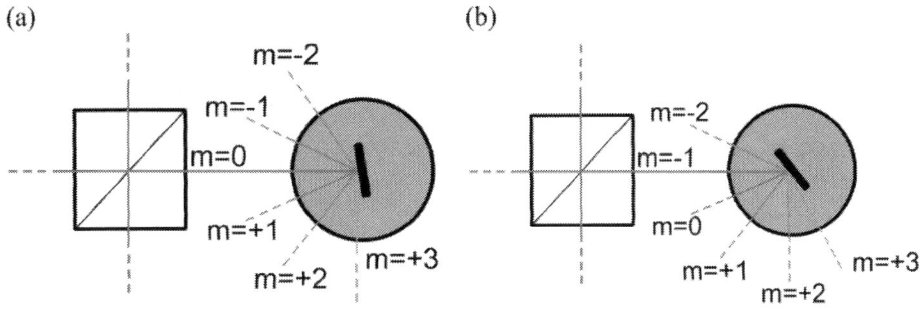


**Figure 5.** Littrow configurations. (a) Zero order. (b) Negative first order.

### *2.1. Zeiss rotary table*
Looking at the equation 1, it can be seen that the rotary table is a critical element in the design. It must be good enough to avoid a high influence in pitch uncertainty and tends to be quite an expensive item. As Inmetro/Lamed already had a high quality rotary table with its CMM, Zeiss RT-05, some tests were carried out in order to determine if it had suitable accuracy, zero repeatability with and without hysteresis consideration. The RT-05 was analyzed by mounting an autocollimator Taylor-Hobson DA-80 with a 12 faces Leitz polygon to measure the rotary angle deviations.

The first experiment was made searching 30 degrees relative positions completing a turn. To find the accuracy of the table, each RT-05 deviation of the relative angle was subtracted by the DA-80 value and corrected by the polygon calibration. The accuracy found was 1 second.

A second experiment was made to analyze the zero repeatability of RT-05, disregarding hysteresis. The RT-05 was triggered to reach a position closer to zero always from +10 minutes to zero, repeating the movement 10 times, taking the difference between the RT-05 deviations from zero to that of the DA-80. The result was a repeatability of 0.1 second for the zero position.

Finally, the last experiment was made to analyze the zero repeat of RT-05 with hysteresis. This experiment was similar to the previously described but the RT-05 was trigger from +10 minutes to zero and sequentially from -10 minutes to zero. The result was a repeatability of 0.3 second in zero position. The results for these experiments revealed an excellent rotary table whose characteristics fit the diffratometry purposes, as will be analyzed next.

### *2.2. Detection path*
Separating the incident light from the retro-diffracted beam, a beam splitter is used to redirect the retro-diffracted beam to an orthogonal direction. Using a mirror and a collimation lens, the retro-diffracted beam is redirected again to be aligned and focused into the CCD camera as a spot image to maximize its resolution.

### *2.3. Optical isolator*
Part of the retro-diffracted beam might return to the laser causing distortions on it what can prejudice the pitch result. The stabilized laser can be unlocked and the wavelength may be changed [10]. In order to eliminate the influence of the retro-diffracted beam effect on laser wavelength, an optical isolator can be used. These optical isolators are composed by waveplates, which turn the beam in two orthogonal polarizations, and polarizers, which can eliminate one of these orthogonal components permitting that only the component aligned with the polarizer goes through. When the beam returns, it will restore linear polarizations but blocked by a polarizer with orthogonal polarization [11].

### *2.4. Intensity controller*
The retro-diffracted beams have less intensity at higher orders. The laser power and image spot brightness are designed so that the weakest order can be detected. However, sometimes the first orders are so bright that the diffraction beam saturates the CCD sensor. An intensity control must be added in order to cut the power of the brighter orders.

### *2.5. Spatial filter*
The spatial filter reduces the wavefront distortions of the incident beam and a lens with an iris is used to collimate the beam and control the spot size (sample area) on the grating.

## 3. Discussion
This part of the paper examines the estimation of pitch uncertainty. This estimation was made consistent with the GUM [12]. Considering the equation 1, the two parameters that have influence on the pitch average are the wavelength of the laser and the angle of grating relative to the zero order

(m = 0). The uncertainty of pitch will be influenced by laser wavelength and the relative angle position, both based on uncertainty of the instruments used for these purposes.

Analyzing the diffractometer schematic drawing (figure 3), optical components have an influence on pitch as well as wavelength and angle. All optical components have imperfections on construction that can disturb the main signal. Some can be analyzed to have a common mode error. The images captured on CCD camera have the same distortion to every diffracted order because each passes through the same optical elements in the same area and in the same direction. Thus, optical elements influence can be removed in a relative analysis as a main pattern response.

The alignment of the system and ambient interferences are the other important influences in the uncertainty of the measured pitch. Basically, the alignment should influence the relative angular position of the grating and the retro-diffracted beam position on CCD camera image. To make the errors due to alignment negligible, a good verification has to be performed to observe and minimize the deviation of the spot position between each order of retro-diffracted beam on CCD camera. Ambient interferences act on all elements of the system and can be modeled as random errors and must be considered in the uncertainty analysis.

The other influences, such as refractive index of air, sample temperature, thermal expansion, humidity, atmospheric pressure and air temperature, were considered for the calculus of the expanded uncertainty estimation (table 1), using ambient values of laboratory standard conditions. The misalignment of the grating is another influence that can interfere with pitch measurement. It is very difficult to measure, but many of these sources are negligible or are in common mode [10]. They are considered negligible in this work.

The estimated result is of the order of picometers for the diffraction first order of a 700 nm pitch sample, which is an excellent result. This is consistent with an international comparison of one dimen-

**Table 1.** Estimated uncertainty budget (order = 1, P = 700 nm)

| Influence | Standard uncertainty | Sensitivity coefficient | Contribution to overall combined standard uncertainty |
|---|---|---|---|
| Laser vacuum wavelength | 8.08 X $10^{-8}$ μm | $P/\lambda_0$ | Negligible |
| Refractive index of air | 4.8 X $10^{-6}$ | $P/n_{air}$ | 3.36 pm |
| Angle traceability | 0.420 second | $P\cot\theta$ | 2.82 pm |
| Angle resolution | 0.578 second | $P\cot\theta$ | 3.86 pm |
| Angle repeatability | 0.382 second | $P\cot\theta$ | 2.56 pm |
| Sample temperature | 1.2 °C | $-\alpha P$ | 0.80 pm |
| Thermal expansion coefficient | 4 X $10^{-6}$ °C$^{-1}$ | $(20-T)P$ | 2.80 pm |
| Humidity | 161.936 Pa | $-\frac{\partial n_{air}}{\partial f}P/n_{air}$ | 0.04 pm |
| Atmosphere pressure | 615.788 Pa | $-\frac{\partial n_{air}}{\partial p}P/n_{air}$ | 1.16 pm |
| Air temperature | 0.578 °C | $-\frac{\partial n_{air}}{\partial T}P/n_{air}$ | 0.76 pm |
| **Combined standard uncertainty** | ---- | ---- | **7.15 pm** |
| **Expanded uncertainty (k=2)** | ---- | ---- | **14 pm** |

sional (1D) grating pitch [13]. As this result is an estimative for just one order, optical elements with no imperfection and standard environmental characteristics of laboratory calibration, it can be expected a larger uncertainty value in a real situation. It can be expected a tens of picometers order variation in uncertainty results with a good manufactured grating.

## 4. Conclusion

The design of a diffractometer involves lot of research and analysis before obtaining an acceptable system. The rotary table, an important part of diffractometer, was tested and revealed a good instrument for this application. The estimation of uncertainty of the pitch measurement was 14 pm for the first order of a 700 nm pitch sample. This value is consistent with the results of an international comparison of one-dimensional (1D) grating pitch [13]. The next steps of the project are the acquisition of all the necessary elements, the diffractometer assembly and then the preliminary tests.

## References

[1] EURAMET 2012 Roadmap Dimensional Metrology *Science and Technology Roadmaps for Metrology - Foresight Reference Document of the Technical Committes of EURAMET e.V.* Draft Update, Retrieved Mar 14, 2014 from http://www.euramet.org/fileadmin/docs/Publications/roadmaps/EURAMET_Science_and_Technology_Roadmaps_for_Metrology.pdf

[2] Lega X C 2012 Model-Based Optical Metrology *Optical Imaging and Metrology: Advanced Technologies* ed W Osten and N Reingand (Weinheim: Wiley-VCH) chapter 13 pp 283–302

[3] Bodermann B, Buhr E, Li Z and Bosse H 2012 Quantitative Optical Microscopy at the Nanoscale: New Developments and Comparisons *Optical Imaging and Metrology: Advanced Technologies* ed W Osten and N Reingand (Weinheim: Wiley-VCH) chapter 12 pp 255–282

[4] University of Eastern Finland n.d. Wave-optical engineering, Department of Computer Science and Statistics, Department of Physics and Mathematics, Retrieved Mar 17, 2014 from http://www.ifc.joensuu.fi/index.php?page=wo

[5] ARDIC INSTRUMENTS n.d. P100 – AFM with CCD portfolio, Retrieved Mar 17, 2014 from http://portfolios.scad.edu/gallery/ARDIC-P100/12316571

[6] Leach R K, Claverley J, Giusca C, Jones C W, Nimishakavi L, Sun W, Tedaldi M and Yacoot A 2012 Advances in engineering nanometrology at the National Physical Laboratory *Meas. Sci. Technol.* **23** 074002

[7] METAS 2015 Measurement Services: Laboratory for Length, Nano- and Microtechnology, Retrieved Feb 6, 2015 from http://www.metas.ch/dam/data/metas/Fachbereiche/Laenge_Nano-_und_Mikrotechnik/Preisliste_5111_E.pdf

[8] Buhr E, Michaelis W, Diener A and Mirandé W 2007 Multi-wavelength VIS/UV optical diffractometer for high-accuracy calibration of nano-scale pitch standards *Meas. Sci. Technol.* **18** pp 667-674

[9] Meli F, Thalmann R and Blattner P 1999 High precision pitch calibration of gratings using laser diffractometry *Proc. of the 1st Int. Conf. on Precision Engineering and Nanotechnology (Bremen)* vol 2, ed P PcKeown *et al* pp 252-255

[10] Pekelsky J R, Eves B J, Nistico P R and Decker J E 2007 Imaging laser diffractometer for traceable grating pitch calibration *Meas. Sci. Technol.* **28** pp 375-383

[11] Williamson R 2005 Polarization Optics Tutorial: Polarizers, Waveplates, Rotators and Lyot Filters, Retrieved Mar 13, 2014 from http://www.ray-optics.com/Polarization_Tutorial.pdf

[12] JOINT COMMITTEE FOR GUIDES IN METROLOGY (JCGM) 2008 Evaluation of measurement data – Guide to the expression of uncertainty in measurement *JCGM* **100**

[13] Decker J E, Buhr E, Diener A, Eves B, Kueng A, Meli F, Pekelsky J R, Pan S-P and Yao B-C 2009 Report on an international comparison of one-dimensional (1D) grating pitch *Metrologia* **48** 04001

Journal of Physics: Conference Series **648** (2015) 012014 doi:10.1088/1742-6596/648/1/012014

# Cox methodology applied to interlaboratory comparison program for flow measurement

**N M Taira[1], O S Yoshida[1], M R Avelino[2,3]**
[1] Institute for Technological Research of São Paulo, Av. Prof. Almeida Prado, 532, São Paulo, SP, 05508-901, Brazil
[3] National Institute of Metrology, Quality and Technology - INMETRO, Xerém, Brazil
[4] State University of Rio de Janeiro, UERJ, Rio de Janeiro, Brazil

E-mail: nmtaira@ipt.br, olga@ipt.br, mravelino@inmetro.gov.br, Mila@UERJ.br

**Abstract.** This paper presents an application of the methodology developed by Cox in Interlaboratory Comparison on water flow measurement. As no reference laboratory wouldn´t be chosen, the comparison was carried out using a reference value by consensus estimated from measurements reported by participating laboratories.

## 1. Introduction

In Brazil, 2008, the Technical Committee in Flow Measurements – TC-13 establishes the first Interlaboratory Comparison (ILC) Program for water flow measurement. Due to the lack of proficiency of laboratories, on flow meter calibration, to meet the accreditation requirements as in NIT-DICLA-026 and in ISO 17025, this intercomparison program was the one of the key actions of the Committee.

In accordance with the General Coordination for Accreditation (CGRE) – NIT-DICLA-042, a technical subcommittee was created involving representatives of the seven laboratories participating of the Interlaboratory Comparison Program. The laboratories were accredited by CGRE in the field of liquid flow and applications. One of the first actions of the Subcommittee was to produce the Protocol of the ILC, whose main propose was to harmonize the information and establish the requirements and procedures to be complied with by the participating laboratories. The Protocol, following the instructions of DOQ-CGCRE-005 – Guidelines for Interlaboratory Comparisons, review of February 1st, 2010, recommended the nomination of a reference laboratory to provide the reference value to be compared with declared values by each participant laboratory. The ILC selected two meters type mass Coriolis as travelling standards, and fortunately, one of them presented reasonable stability during the entire program, which lasted over 18 months. However, the ILC results were not well accepted because of unexpected rejection of two laboratories, which raised the possibility of not reliability of the reference value.

After the document DOQ-CGCRE-005 was withdrawn in 2012, the TC-13 held a second edition of ILC, this time without the nomination of a reference laboratory to indicate the reference value, but with the reference value estimated from the values of each participant laboratory. The reasons for this decision were due to low maturity and experience of participant's laboratories, besides, the Cox [1] procedures for the statistical analysis of comparison measurements could overcome the difficulties on comparing data with large range on uncertainties and from laboratories with high discrepancies in some flow rates.

The section 2 of this paper presents the Cox methodology (Procedure A) for calculating the reference value as the weighted mean together with tests to check the consistency of a comparison data of four laboratories, that fails because of discrepancies within laboratories at low flow rates that could not be removed from data set. As the consistency test failed, the Cox-methodology (Procedure B) was applied using as a reference value the median since it is robust for the inconsistencies not removed from the data set. Section 3 presents merits of the Cox methodology applied on the comparison data of seven participating laboratories in ILC since the first edition of the program.

## 2. Calculation methodology to flowrate interlaboratorial comparison programs

The most preferred performance statistic for ILC is the $E_N$ number, as recommended by the extinct DOQ-CGCRE-005-2005 and by the standard ABNT NBR ISO/IEC 17043:2011 [2]. The definition of $E_N$ is presented in equation (1),

$$E_{N_i} = \left| \frac{(value\ declared\ by\ laboratory\ i) - (reference\ value\ )}{\sqrt{U_{lab_i}^2 + U_{reference}^2}} \right| \qquad (1)$$

where

$U_{lab_i}$ Expanded uncertainty of laboratory *i*,

$U_{reference}$ Expanded uncertainty of the reference value.

The criteria limit is 1, so the measurement is considered consistent if EN<1 or inconsistent for $E_N$>1. Cox [1] presents procedures developed to obtain the key reference value (KCRV) for comparisons involving key laboratories of National Institutes of Metrology (NMI). The procedure proposed by Cox has been applied on many Interlaboratory Comparison Programs, and it is well documented in metrology literature: as in Mikan [3] and in Arias et al. [4]. Although the Cox methodology was developed for comparison of key laboratories it is also very suitable for common laboratory comparisons through efficient use of more sophisticated statistical tools.

### *2.1. "Procedure A" proposed by Cox*

The Procedure A estimates the reference value by the mean of the values declared by the laboratories weighted by the their respective standard uncertainties, according to equation (2),

$$y = \frac{\frac{x_1}{u_{x_1}^2} + \frac{x_2}{u_{x_2}^2} + \cdots + \frac{x_n}{u_{x_n}^2}}{\frac{1}{u_{x_1}^2} + \frac{1}{u_{x_2}^2} + \cdots + \frac{1}{u_{x_n}^2}} \qquad (2)$$

where $x_1, x_2, \cdots, x_n$ are the errors of the meter at a nominal flow rate, as declared by the laboratories $1, 2, \cdots, n$ , and $u_{x_1}, u_{x_2}, \cdots, u_{x_n}$ are the uncertainties of the errors reported by the respective laboratories $1, 2, \cdots, n$.

The uncertainty of the reference value is obtained by equation(3)

$$\frac{1}{u_y^2} = \frac{1}{u_{x_1}^2} + \frac{1}{u_{x_2}^2} + \cdots + \frac{1}{u_{x_n}^2} \qquad (3)$$

The data consistency test, equation (4-5), or chi-square test, verifies the validity of the dataset to produce a reliable comparison reference value (CRV),

$$Pr\{\chi^2(\nu) > \chi_{obs}^2\} < 0.05 \qquad (4)$$

where,

$$\chi^2_{obs} = \frac{(x_1 - y)^2}{u^2_{x_1}} + \frac{(x_2 - y)^2}{u^2_{x_2}} + \cdots + \frac{(x_n - y)^2}{u^2_{x_n}}. \tag{5}$$

For the $\chi^2$ test it is necessary that all measurements are $N(x_i\,, u^2_{x_i})$.

Having validated the data set, the next step is to determine the discrepancies between the error stated by the laboratories relative to the reference error, $\mathrm{y} = \mathrm{x}_{\mathrm{ref}}$, ie., $\mathrm{d}_\mathrm{i} = \mathrm{x}_\mathrm{i} - \mathrm{x}_{\mathrm{ref}}$. The acceptance criteria is presented in equation (6) or equation (7),

$$|d_i| \leq 2u(d_i) \quad (6) \qquad \text{or} \qquad E_N = \left|\frac{d_i}{2u(d_i)}\right| \leq 1 \quad (7)$$

where $u(d_i) = \sqrt{u^2_{x_i} - u^2_{x_{ref}}}$.

The four participating laboratories calibrated of the standard meter in ten points in the range 15 to 3000 L/h. Figure 1 shows the flow meters calibration curves and the weighted mean curve as the reference value (CRV) curve.

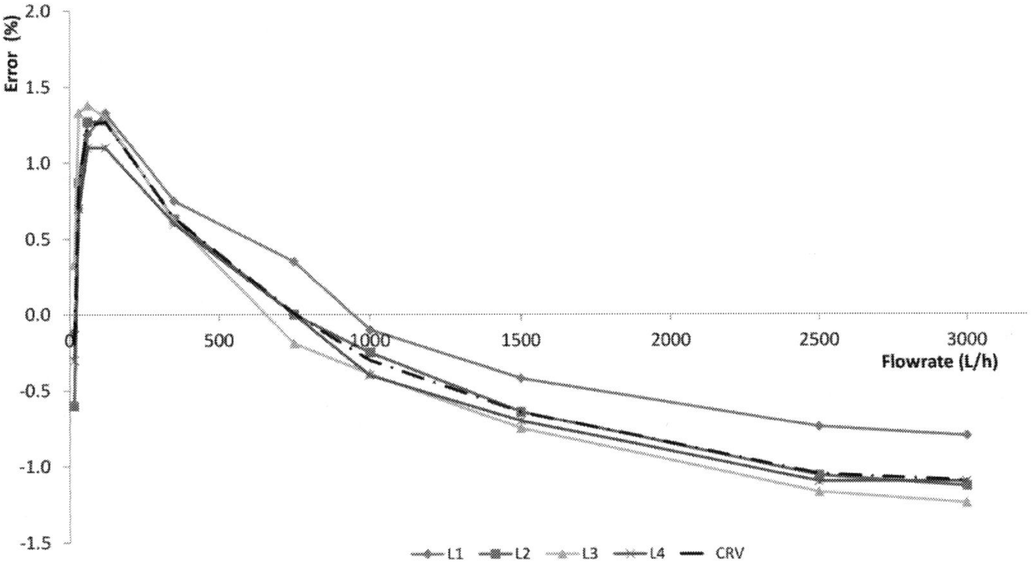


**Figure 1.** Flow meters calibration curves by four laboratories and the reference curve (CRV)

Figure 2 presents the curves (EN). It is clear from the Figure 2 that, according to the criteria defined by the equation 7, results obtained for L1 laboratory are not consistent for flow rates above 750 l/h, and the laboratories L3 and L4 are not consistent on others points.

Table 1 shows, for each flow rate, the reference values (CRV) with their uncertainties, and the results of data consistency tests in (4). It can be observed in Table 1 (a) that only four points could be used for comparison of the results. However, as recommended in Cox (2002), the data which could have generated the inconsistencies may be withdrawn. Thus, Table 1 (b) shows the results obtained without taking into account the laboratory dataset **L1**, which had the highest amount of (EN) points above the acceptable limit (Fig.2).

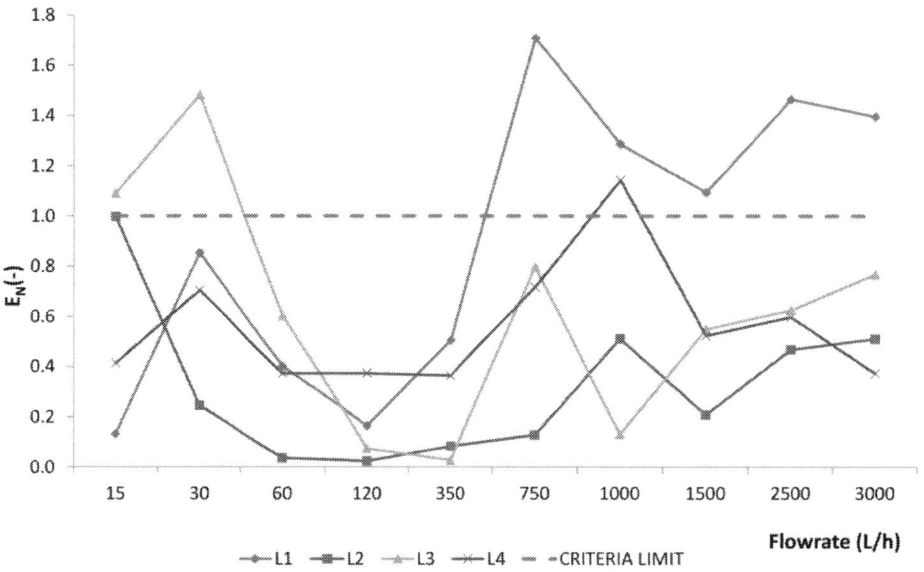


**Figure 2.** $E_N$ values of four participating laboratories obtained by Procedure **A**.

**Table 1.** Reference values and data consistency test, according to procedure A.

| Flowrate (L/h) | CRV (%) | standard uncertainty of CRV (%) | $\chi^2_{obs}$ |
|---|---|---|---|
| 15 | -0.122 | 0.071 | inconsistent |
| 30 | 0.988 | 0.098 | inconsistent |
| 60 | 1.288 | 0.079 | ok |
| 120 | 1.289 | 0.075 | ok |
| 350 | 0.613 | 0.034 | ok |
| 750 | 0.027 | 0.033 | inconsistent |
| 1000 | -0.357 | 0.033 | inconsistent |
| 1500 | -0.598 | 0.058 | ok |
| 2500 | -0.982 | 0.056 | inconsistent |

**(a)** All four laboratories

| Flowrate (L/h) | CRV (%) | standard uncertainty of CRV (%) | $\chi^2_{obs}$ |
|---|---|---|---|
| 15 | -0.152 | 0.133 | inconsistent |
| 30 | 1.085 | 0.113 | inconsistent |
| 60 | 1.329 | 0.095 | ok |
| 120 | 1.274 | 0.088 | ok |
| 350 | 0.604 | 0.035 | ok |
| 750 | -0.012 | 0.035 | ok |
| 1000 | -0.384 | 0.034 | ok |
| 1500 | -0.688 | 0.071 | ok |
| 2500 | -1.094 | 0.068 | ok |

**(b)** Without laboratory L1

Although discrepancies still remain at the two lowest flowrates of the calibration range, neglect them is not a reasonable decision, once it is expected that flow meters have low performance at low flow rates. For this case, Procedure **A** is not made viable, even by following the recommendations in Cox (2002), and one may proceed to the application Procedure **B**, which is robust for datasets with some discrepant measurements.

### *2.2. "Procedure B" proposed by Cox*

Procedure B associates each value declared by the participating laboratory to the normal distribution and to the mean $x_i$ and standard deviation. Samples are generated from the distribution by simulation and their median, the mean of the medians is the reference value.

Figure 3 shows examples of histograms of the values of the median estimated from the simulation ($10^6$ samples) of results of four laboratories and their distributions reference value calculated according to the Procedure B.

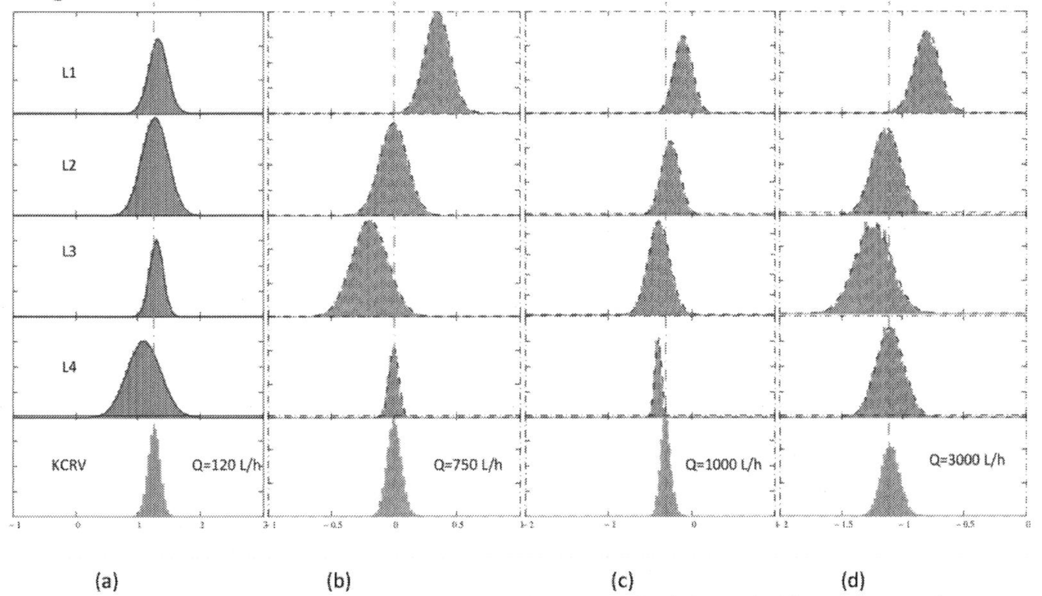


**Figure 3.** Histograms obtained using the Cox B methodology for four laboratories

Thus, one can define a new equation for the normalized/standard error $E_N$ according to the equation (7), defined as the degree of equivalence according to procedure B. Cox, which is calculated according to ISO GUM [5] also adopted by Arias et al. [4], and also according to Appendix B of Cox(2002). In his work, Cox suggests to estimate the lower confidence interval for each value of $d_i$. Table 2 presents the results of the ILC in a volume of 100 mL artifact conducted within the CIPM, where are presented the *DoE* calculated values and the calculated values according to the computational tools called *Cox_B* and *Cox_B_full*, which are described in item 2.3.

In Table 2 it is possible to notice the adherence of the *DoE* declared values and the ones calculated with the developed computational tools. Recalling that those are not exact values once they are calculated by using the simulation date.

**Table 2.** Application of Cox methodology for CIPM key comparisons.

| NMI | declared values | | *DoE (Degree of Equivalence)* | | |
|---|---|---|---|---|---|
| | $x_i$ (mL) | $x_i$ (mL) | Cox_Arias* | Cox_B | Cox_B_full |
| CENAM | 99.89350 | 0.00077 | 0.44 | 0.44 | 0.45 |
| NRC | 99.89780 | 0.00080 | 1.80 | 1.81 | 1.87 |
| SP | 99.89500 | 0.00160 | 0.31 | 0.32 | 0.32 |
| IMGC | 99.89300 | 0.00083 | 0.67 | 0.65 | 0.70 |
| NMIA | 99.89550 | 0.00110 | 0.65 | 0.62 | 0.66 |
| INMETRO | 99.89290 | 0.00061 | 0.81 | 0.81 | 0.87 |
| KCRV | 99.8942 | 0.00060 | | 99.8942 | 0.00060 |

*Adjusted from de Arias et al. [4]

### 2.3. Computational Tools

For the purposes of the Cox Procedure B in an ILC of flow calibrations hundreds of millions of data are computed. For this reason, the use of basic computer tools has increasingly become inappropriate. Due to the considerable processing time and computational efforts the use of spreadsheet has shown itself inconvenient when over $10^4$ samples were simulated. Thus, both procedures **A** and **B**, as proposed by Cox (2002), are developed for *Mathcad®* e *VBA Excel®* programming environments. Additionally, a Supplement *CSharp Excel (Add-Ins)* was specifically developed to include functions of *Cox_A( )*; *Cox_B( )* e *Cox_B_full( )*. This implementation is compatible for *Excel® Office®* package, in versions 97, 2003, 2007 and 2010.

The *Cox_A* function calculates the comparison reference value (CRV), mean differences ($d_i$) and the degrees of equivalence (*DoE*) from the values reported by participating laboratories of a ILC according to the procedure proposed by Cox, and described in Section 2.1.

The *Cox_B* function calculates the comparison reference value of CRV by the "median method" calculated from the Monte Carlo simulation of M random values of the inverse of the normal cumulative probability distribution. The parameters taken into account are the declared value (mean) and the uncertainty specified by the each of the participating laboratory. According to Cox (2002), it is recommended to use M = $10^6$ random elements to characterize the values in a population with a distribution similar to the normal.

The *Cox_B* function produces the following result:

$u_{CRV}$ Standard uncertainty of the value of CRV;
$d_i$ calculated average difference for laboratory i relative to the reference value;
$u(d_i)$ uncertainty of the mean difference $d_i$ calculated according to ISO GUM [5];
$DoE$ equivalence degree defined as the magnitude of the ratio $d_i/(2u(d_i))$.

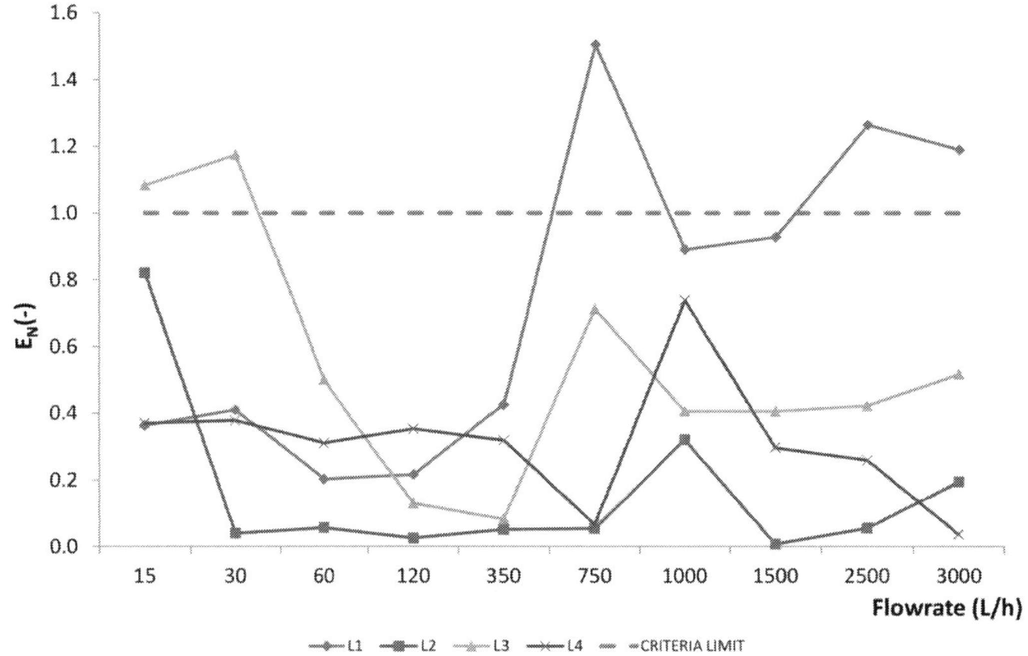


**Figure 4.** $E_N$ values of four participating laboratories obtained by *Cox_**B*** function.

Figure 4 shows the result of ILC with four participating laboratories, by applying the Procedure **B** to the same data of the Figures 1 and 2. It is observed the same trend of $E_N$ for L1 and L3 laboratories, but the laboratory L4 no longer presents results higher than recommended for ILC, as if it was possible to apply Procedure **A**.

The *Cox_B_full* function calculates the reference value in the same way that *Cox_B* function, but $u_{d_i}$ is calculated in accordance with Appendix B of Cox(2002) that estimates the lower confidence interval for each value.

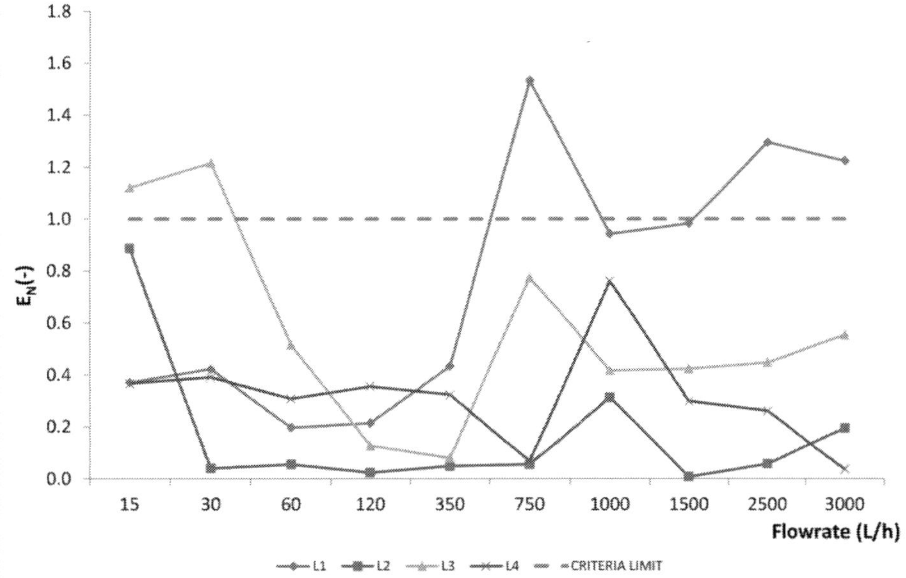


**Figure 5.** $E_N$ values of four participating laboratories obtained by Cox_B_full function.

Figure 5 shows the results obtained by applying the function to *Cox_B_full* to the same set of data relating to Figures 3 and 4.

Comparing the results generated by the three functions described, for the same set of data, it is observed that the *Cox_A* generates more "conservative" results and that the *Cox_B* function generates more "optimistic" results, that is, according to the understanding of the participating laboratories. On the other hand, the *Cox_A_full* function may be considered more "careful", as shown in the *DoE* values, as shown in Table 2.

### 3. Application of the tools in the flowrate interlaboratorial comparison program

Figure 6 shows the result of ILC in mass flowrate with seven participating laboratories and only one laboratory. Results of Laboratory **E** are considered to be unsatisfactory. For this ILC one laboratory was designed to be the reference laboratory, as described in item 1.

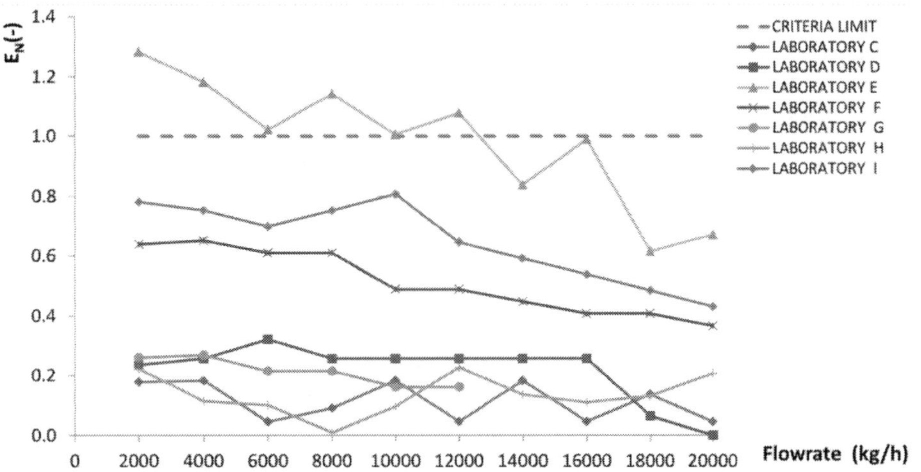


**Figure 6.** Example of $E_N$ values in ILC mass flow with seven participating laboratories

The application of Cox_A function results in inconsistencies in various flowrates. By applying the *Cox_B_full* function it is possible to notice the existence of two groups of laboratories whose artifact calibration curves have different shapes in relation to the CRV reference value, as shown in Figure 7.

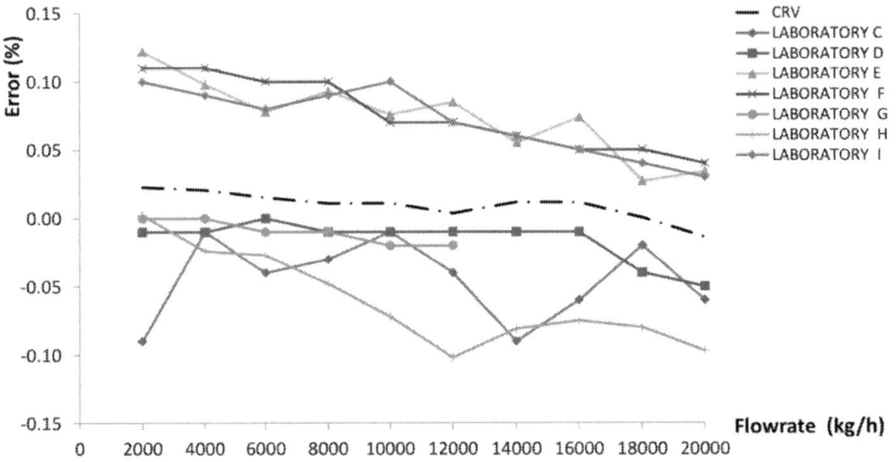


**Figure 7.** Mass flowrate meter calibration curves relative to the CRV obtained thought *Cox_B_full* function in ILC with seven participating laboratories

The application of *Cox_B_full* function generates the (*DoE's*) where all the laboratories showed satisfactory results, including the laboratory **E**. The same applies to the reference laboratory, as shown in Figure 8.

For the case shown in Figure 8 expanded uncertainties reported by participating laboratories ranged from 0.06% to 0.22%, ie, the uncertainties are not "harmonized" and these can affect laboratory results that declared the highest value and to promote the laboratory that declared lower uncertainty values, as can be deducted from equation 2 for the case of Procedure **A**. The same applies to Figure 3 for the case of Procedure **B**.

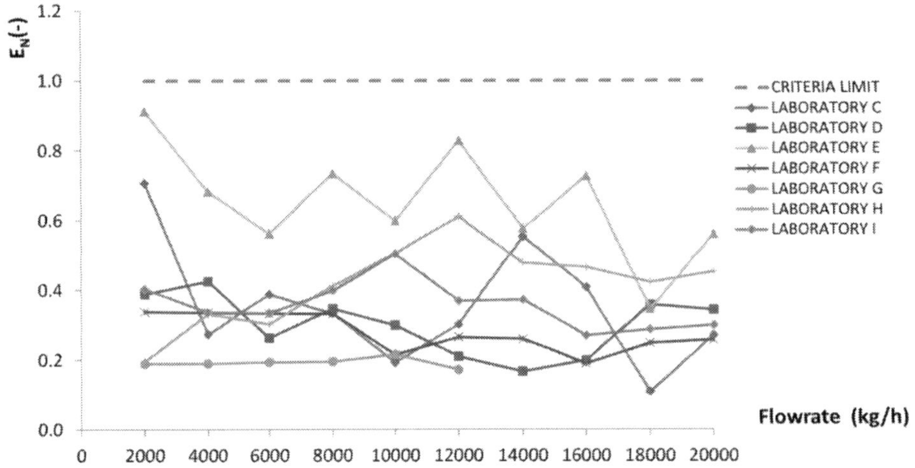


**Figure 8.** Curves of $E_N$ calculated by *Cox_B_full* function in ILC mass flow with seven participating laboratories

Assuming that all laboratories reported an expanded uncertainty uniform and equal to 0.06%, it is found that there will be no change in the form of the error curve as shown in Figure 9. The $E_N$ curves, shown in Figure 10, indicate that four laboratories had unsatisfactory results.

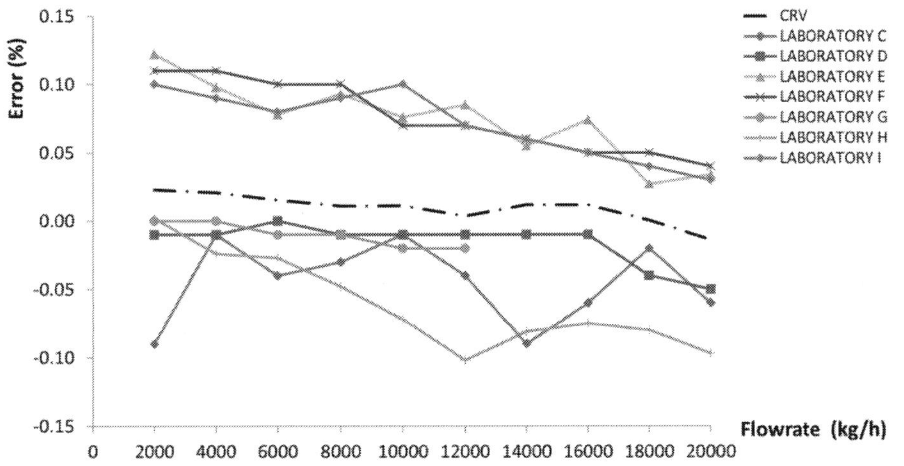


**Figure 9.** Mass flow meter calibration curves relative to the CRV calculated as *Cox_B_full* function in ILC with seven participating laboratories and expanded and uniform uncertainty equal to 0.06%.

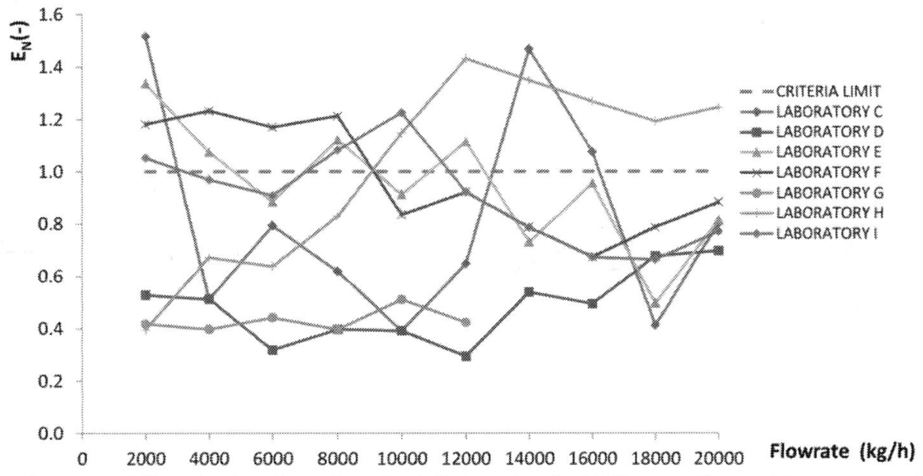


**Figure 10.** Curves of $E_N$ calculated as *Cox_B_full* function in ILC mass flow with seven participating laboratories and expanded and uniform uncertainty equal to 0.06%.

Assuming that all laboratories reported an expanded uncertainty uniform and equal to 0.22%, it is verified from Figure 11 that there are no changes in the form of the error curves. The $E_N$ curves indicate that all laboratories present satisfactory results, as shown in Figure 11, as was expected.

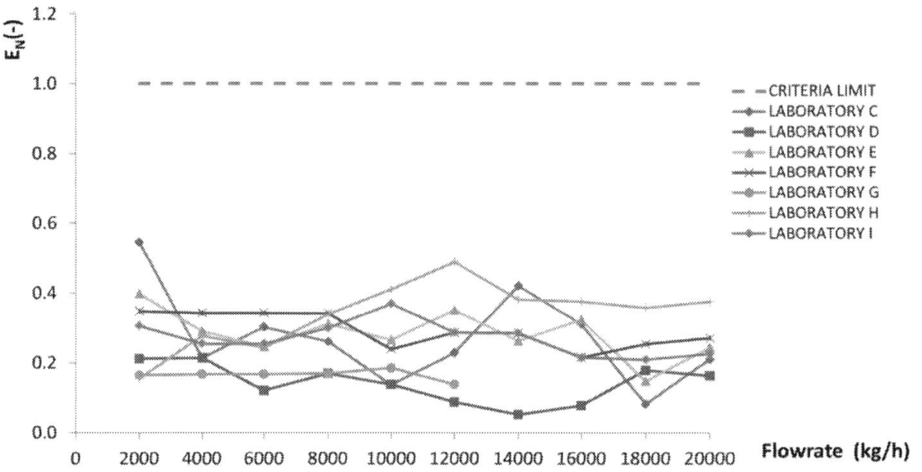


**Figure 11.** Curves of $E_N$ calculated as *Cox_B_full* function in ILC mass flow with seven participating laboratories and expanded and uniform uncertainty equal to 0.22%.

## 4. Conclusion

Although there are several acceptance criteria and calculation methodologies for the ILC could be established by consensus or to be defined by the laboratory accreditation manager, as described in the standard ISO/IEC 17043, there are, on the field of flow measurement in Brazil, a considerable lack of knowledge and common understanding on the measurement uncertainties likely to be reached by the flow meter calibration laboratories.

The Cox methodology was also applied in ILC on water meters with the participation of manufacturer's laboratories, laboratories of water companies and independent laboratories. During the first edition of ILC, several unsatisfactory laboratory performances occurred, resulting largely from the lack of harmonization of the declared uncertainties, which ranged from 0.01% to 0.64%. During the second edition of the ILC, there were a significant progress and an improvement in ensuring the metrological reliability in calibration activities and the water meter verification activities in Brazil, as well as the dissemination of best practices and harmonization of metrological concepts. The application of Cox procedures had a fundamental role on this improvement process.

The application of methodologies Cox proved to be flexible and efficient to promote harmonization of the laboratories performances as the application of procedure **B** would keep on comparison data set the results of discrepant laboratories or laboratories with uncertainties that could not cover their measurements, allowing the learning of the not mature laboratories.

## References

[1] Cox M G, 2002 The evaluation of key comparison data *Metrologia* 39, p.589-595

[2] Associação Brasileira de Normas Técnicas, 2011 *ABNT NBR ISO/IEC 17043: Avaliação de conformidade – Requisitos gerais para ensaios de proficiência.* (Rio de Janeiro: ABNT)

[3] Mikan B and Valenta T, 2009 *Final Report - Draft B, interlaboratory comparison calibration of the turbine gas meter G6500 EURAMET Project No. 1006*, March

[4] Arias R et al., 2005 *CIPM Key Intercomparison volume at 20 L and 100 mL* - CCM.FF-K4, March

[5] International Organization for Standardization, ISO, 1993 Guide to the expression uncertainty in measurement.

# DFA concepts applied in development of accessories for calibration

**A P Silva[1,2], R Suterio[1,2]**

[1] INPE National Institute for Space Research, São José dos Campos, SP, Brazil
[2] ITA Technological Institute of Aeronautics, São José dos Campos, SP, Brazil

E-mail: alberto@lit.inpe.br

**Abstract**. The purpose of this paper is to identify design requirements of a device to be used as an accessory in calibration. In this case study, a simple device to sample atmosphere from humidity generators used to calibrate air humidity meters is developed and it is proposed a model applying Design for Assembly concepts.

## 1. Introduction

The Physical Metrology Laboratory (MTF), part of the National Institute for Space Research (INPE)/Integration and Testing Laboratory (LIT), is responsible for calibrating sensors and equipment used by INPE, private companies, universities and other research institutes. Its area of actuation is acoustics and vibration, temperature, humidity and vacuum.

Calibration of humidity sensors and hygrometers are usually performed through comparison between the indication of the device under calibration and a conventional value indicated by a standard traceable to the International System (S.I.), using a calibration chamber with stable and controlled humidity.

The aim of this paper is to apply some general concepts from Design for Assembly (DFA) methodology to guide the design of an optimized device to be used in humidity calibration, specifically in dew point calibration.

## 2. Design for Assembly

DFA is a systematic method derived from Design for Manufacturing (DFM), in which great improvements come from simplifying the product through minimizing the number of separate parts.

Boothroyd and Dewhurst DFA methodology purposes three criteria to guide the analysis for each product part in the assembly [1]:

a) Does the part have relative movement to the other parts of the product?
b) Does the material have to be different from the other parts already assembled?
c) Does the part have to be separate from others to allow product assembling and disassembling?

Any affirmative answer to those questions indicates that the part shall be separate.

Some of good practices to achieve improvements in product design are [2]:

- Easy handling and assembling of the product due the small number of pieces;
- Modular design and the possibility of product variants from new parts to be integrated in the future;
- Choose the material by function and process;
- Parts that cannot be assembled incorrectly;
- Efficient use of fixation techniques and
- Use of gravity.

At this point, manufacture feasibility is not considered. However, feasible ideas can arise [1].

## 3. Humidity measurement

Humidity measurement can be made by total or partial immersion of the probe or sensor in the environment to be measured, or by collecting a sample of air from this environment through a pipe. Two methods are widely used for generating humidity calibration environment: the use of climatic test chamber and the use of generator dewpoint [3]. For best results, it is recommended to place the sensor as close as possible to the point of measurement. When using a climatic chamber, for instance, the sensor can be placed inside the environment to be measured. Collecting a sample of air from the environment is the most appropriate method in cases in which it is needed to measure the temperature and dew point and the gas is contaminated, with filtering possibility, and when the gas temperature is higher than limit sensor operation [4]. It is also necessary to sample the atmosphere when the dew point generator does not provide a measuring chamber for the sensor to be calibrated. To solve this problem, the flow of air generated can be driven by the item being calibrated and the working standard, enabling the comparison of both [3].

In the calibration of optical dew point hygrometers it is necessary to use a measuring chamber, which will receive the gas sample and should provide the same dew point conditions from the generator. For a measurement of dew point close to the true value, sampled gas shall be conducted from its source to the sensor mirror without distortion of the partial pressure of water vapor [5]. Some dew point meters already have small measurement chambers.

Even if the sensors used in measurement have a very low response time, there may be a delay or failure in the correct measurement of humidity, especially at low values. This is due to several factors that should be considered in sampling gas from the environment to be measured.

Any type of contaminant should be avoided, including sources of water vapor. The lower the humidity to be measured, the most accurate care is demanded with the measurement environment [3]. There are specific recommendations for each measurement range. To minimize contamination, the sensor mirror should stand in the vertical position. If the temperature of dew point of the gas is higher than the room temperature, it shall be provided heating of the sampling line and the sensor in order to avoid condensation of the measurement gas [4].

The number of connections shall be reduced to minimize the leakage of moisture and dead ends should be avoided as they hinder the removal of moisture from the system [3]. The impact of leaks is inversely proportional to the dew point temperature to be measured, being very significant in the negative range [4]. Another source of moisture that should be minimized is the feedback, i.e., the return of water vapor after the sensor into the system. A good practice is the use of long exhaust pipe after the sensors. The choice of materials used in measuring low humidity is also of great importance for the efficiency of the system. Use of hygroscopic materials shall be avoided [3]. Material absorption/desorption characteristcs affect the system as a whole [4]. Stainless steel and nickel are considered virtually impervious, what make them recommended for this application. Comparatively to steel, polytetrafluoroehtylene (PTFE) is very permeable. Although, among the polymeric materials it is suitable for measuring humidity to -20 °C dew point and in some cases even slightly below this value [3]. Most plastics and rubbers are not suitable for measuring any moisture range [4]. Figure 1 shows the performance of various materials in moisture removal in a measurement system.

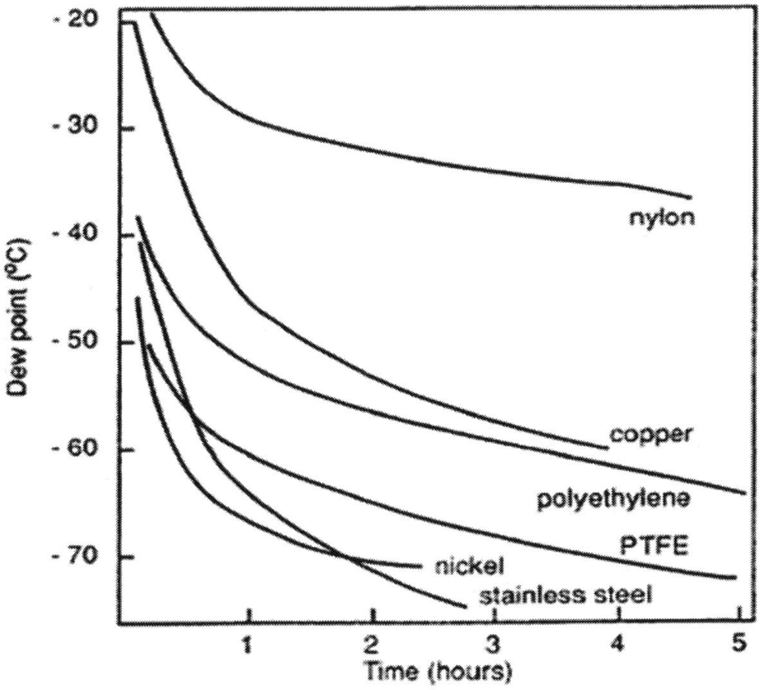


**Figure 1.** Materials performance in moisture removal [1]

## 4. Case study

INPE/LIT has a dew point calibration system (DCS) comprising a dew point generator, a dew point optic hygrometer and a port to carry the calibration gas. The item under calibration shall be exposed to the calibration gas. This was the reason to design a device for sampling the gas and put it in contact with the sensor.

Initially, two sensors used by INPE/LIT as humidity working standards were chosen. The sensors under calibration using the sampling device have different measuring principles, operation and dimensions. The first sensor (Sensor A) is a two stage cooled mirror optical dew point sensor part of a dew point hygrometer model Optidew Vision by Kahn Instruments, Figure 2. The second sensor (Sensor B) is a capacitve humidity sensor model HC2-IC105 by Rotronic, Figure 3.

**Figure 2**. Optical sensor under calibration using the humidity sampling device (www.michell.co.uk)

**Figure 3.** Capacitive sensor under calibration using the humidity sampling device (www.rotronic-usa.com)

**Table 1.** Technical data of sensors under calibration

| Feature | Sensor A | Sensor B |
|---|---|---|
| Sensor type | Optical | Capacitive |
| Dew point working range | -40 a +60 °C | -40 a +60 °C |
| Application | Working standard for humidity calibrations | Humidity monitor for EMI/EMC tests |
| Sensor length | 63 mm | 100 mm |
| Sensor diameter | 28 mm | 15 mm |
| Mechanical connection | M36x1.5-6g | Fitting |
| Minimum immersion depth for measuring | 74 mm | 39 mm |

Based on the application, three attributes were considered more relevant in this case: efficiency, ease of operation and adaptability. This decision leaded to a list of desirable characteristics, as shown in Figure 4.

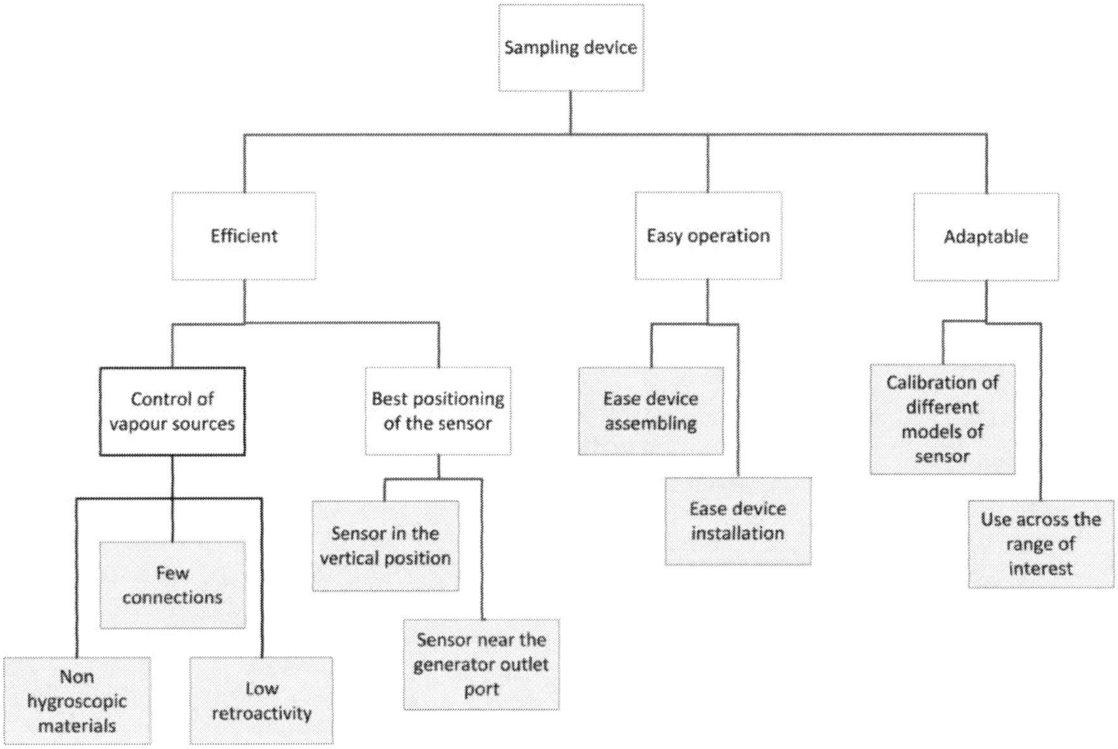


**Figure 4.** Desirable characteristics for the humidity sampling device

The efficiency of the device is strongly linked to the prevention of unwanted sources of water vapor as points of accumulation of moisture absorbed or adsorbed and entry points for moisture. Once these sources, in addition to increasing the settling time of the system, can introduce errors and increase the measurement uncertainty, the proposal is to minimize its occurrence.

From the characteristics of the sensors to be calibrated, the moisture range defined in this work and the desirable characteristics of the device, project requirements were identified:

- The device must have only non-hygroscopic materials.

Journal of Physics: Conference Series **648** (2015) 012015 doi:10.1088/1742-6596/648/1/012015

- The device must have connections only at the points where there is input or output of gas and where the sensor is installed.
- The device must allow entry and exit of gas.
- The device should allow the installation of a moisture sensor in the horizontal direction.
- The device must have a pipeline to carry gas from the generator dew point until its entry.
- The device must have a duct in the output.
- The device should allow installation of sensors with diameter up to 28 mm.
- The minimum immersion depth of the sensor in the measuring chamber shall be 39 mm.
- The device must meet the range of -40 to +20 ° C dew point temperature.

Starting with the choice of materials, the option was for the use of stainless steel and PTFE. The use of PTFE was admitted as possible because its characteristic absorption and desorption of moisture is at the limit to the range of this work.

The suggested initial setting is shown in Figure 5. In this configuration the device comprises:

- 01 measuring chamber with a gas inlet connection with thread, a connection for gas outlet with thread and with an open face to allow installation of different lids.
- 01 lid with hole for Sensor A.
- 01 lid with hole for Sensor B.
- 01 pipe for connection to the dewpoint generator, threaded at both ends.
- 01 pipe for connection to the output of the measuring chamber to avoid feedback.
- 04 screws on the lids.

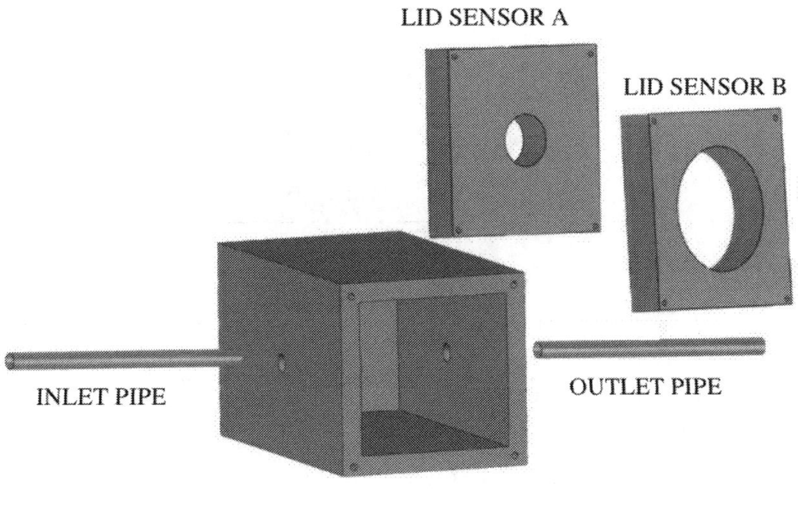


**Figure 5.** Preliminary design

The fastening of the lids in this configuration is a problem for the efficiency of the system because of the sealing and the assembling requires a tool and alignment. The number of connections can also be a problem.

Applying the criteria proposed in Boothroyd Dewhurst DFA method it was observed that some parts could be unified. Based on the criteria presented in Section 2 the analysis is presented in Table 2.

When applying DFA it was noticed that a change in the concept of the chamber could make it possible to remove the lid dedicated to Sensor A. Therefore, the interior of the measuring chamber was based on the dimensions of the Sensor A, which has a thread on its body that enables easy connection to the calibration medium, as shown in Figure 2.

Journal of Physics: Conference Series **648** (2015) 012015 doi:10.1088/1742-6596/648/1/012015

The lid for the Sensor B can not be ruled out. However, as this sensor has a smaller diameter than the diameter of the sensor A, it was possible to alter the geometry and the way of fastening this lid. Making it circular with the same external diameter and thread from Sensor A, the four holes and screws could be removed.

**Table 2.** Application of criteria to minimize parts

| Part | a) relative movement | b) different material | c) must be separate |
|---|---|---|---|
| Measurement chamber | no | no | yes |
| Lid - Sensor A | no | no | no |
| Lid - Sensor B | no | no | yes |
| Inlet pipe | no | no | yes |
| Outlet pipe | no | no | no |
| Screws | no | no | yes |

The inlet pipe for connection to the dew point generator must remain separate as it can be used for calibration of sensors that have their own measuring chambers. Furthermore, it can be replaced anytime by another tube of the same material or different material, increasing the flexibility of the system.

The outlet pipe has to minimize the retroactivity single function and can be part of the measuring chamber, reducing one step in assembly and disassembly and also reducing connection in the system.

The new configuration is shown in Figure 6.

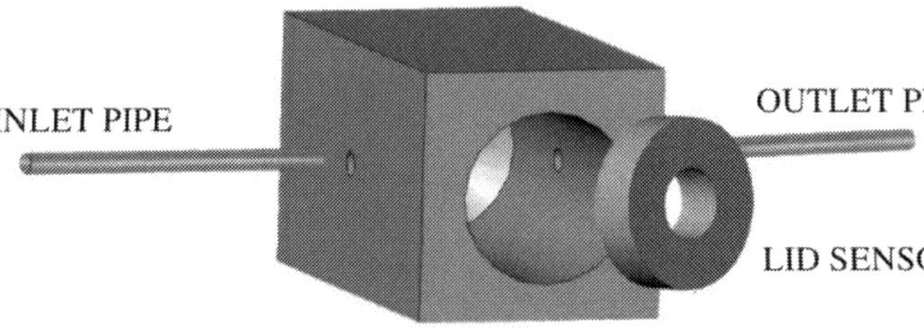


**Figure 6.** New design after DFA

Modifications after applying the DFA resulted a device with the following parts:

- 01 measuring chamber with a gas inlet connection with thread, a welded gas outlet pipe to minimize feedback, a hole with a M36x1.5 thread for installation of sensor or lid.
- 01 lid with external M36x1.5 thread and drilling for Sensor B.
- 01 inlet pipe for connection to the dew point generator, threaded at both ends.

The external geometry of the sampling device has been chosen to dispense the use of supports which give stability to the assembly.

## 5. Conclusion

Some requirements for the sampling device are quite obvious, because they are identified with the characteristics of the sensors and the working range defined by the user. However, after studying the factors that can influence the performance of the calibration system as a whole, it has been found necessary to include requirements disregarded in the initial analysis.

Although DFA aim is to simplify the assembly of products, in order to simplify manufacturing, its application in this study guided the design of the sampling device to meet the identified metrological requirements and improve its operational ease.

The criteria proposed by Boothroyd and Dewhurst DFA method were effective in decreasing the parties that may affect device performance, particularly in the reduction of points accumulation, leak or desorption of water vapor. The proposed model after applying DFA meets the requirements of the project and should be submitted in the next step of the manufacturing feasibility analysis, which was not considered in this work.

There is a significant indication that DFA can be applied in others areas of metrology to help in the design of optimized accessories for calibration.

**References**

[1] Boothroyd G 1996 Design for Manufacture and Assembly: The Boothroyd and Dewhurst Experience *Design for X: Concurrent Engineering Imperatives* ed Huang G Q (London: Chapman & Hall)

[2] Leaney P G 1996 Case Experience with Hitachi, Lucas and Boothroyd-Dewhurst DFA Methods *Design for X: Concurrent Engineering Imperatives* ed Huang G Q (London: Chapman & Hall)

[3] National Physical Laboratory 1996*A Guide to the Measurement of Humidity* (London: The Institute of Measurement and Control)

[4] Cortina V B 1985 Sampling Systems for Chilled Mirror Dew Point Hygrometers *Moisture and Humidity Measurement and Control in Science and Industry: Proc. of the 1985 International Symposium on Humidity and Moisture* (Washington D.C., USA, 1985) ed Instrument Society of America (Durham: Instrument Society of America) pp 849-51

[5] Paine L C, Farrah H R 1965 Design and Applications of High-Performance Dew-Point Hygrometers *Humidity and moisture: measurement and control in science and industry* ed Wexler A (New York: Reinhold Publishing)

# Flow meter calibration by volumetric method and by weighing method using an innovative system

**V YAibe, M H G Aquino, E C C Farias, P C Gabriel**
Inmetro, Duque de Caxias, Brazil

E-mail: pcgabriel@inmetro.gov.br

**Abstract.** A flow meter can be calibrated by different methods. To obtain minors measurements uncertainties, the laboratories use the weighing method; however, this method does not provide a simple operation, a specific apparatus has been developed for this calibration. Another approach is the calibration by volumetric method, using a tank prover as standard. This paper show the results obtained by this two different methods for the same flow meter. Key words: flow meter, weighing method, volumetric method, measurement uncertainty, innovation.

## 1. Introduction

The flow meter calibration occurs by different methods. A calibration by gravimetric method delivers the lowest measurement uncertainty, however the complexity of the method is greater. In a metrology laboratory the volumetric method is frequently used, however this method does not deliver the best measurement uncertainty.

The approach of this paper aims to compare the calibration of a same flow meter by the weighing method and volumetric method, in different layouts.

For this job, an oval gear flow meter was used, with range of 50 L/min to 500 L/min. Water cistern was used as calibration fluid.

## 2. Calibration method

The meter was calibrated in first moment by weighing method. For this method were used two equal balances with capacity of 300 kg and 2 g resolution [Fig. 1]. The balances were calibrated before and after of meter calibration. One 350 L capacity container was positioned on each balance. The totalized volume by flow meter was calculated as the difference between full mass and empty mass on the containers. A new device was developed for the totalized volume on the flow meter be maintained constant.

For volumetric method a tank prover was used as standard [Fig. 2] to calibrate the flow meter. The tank prover was calibrated using weighing method previously by Liquid flow laboratory (Laliq) of Inmetro internal standard. The equation 1 gives the mathematic model to determine the meter factor.

$$F_m = \frac{V_r}{V_m} \cdot \left(1 - \alpha_m \cdot \left(T_{Lm} - T_c\right)\right) + \delta F_m \qquad (1)$$

Journal of Physics: Conference Series **648** (2015) 012016 doi:10.1088/1742-6596/648/1/012016

Where:

Fm – Meter factor for totalized volume in flow meter (dimensionless)

Vr – Calculated volume by weighing method or volumetric method, in dm3.

Vm – Totalized Volume by flow meter (pulse number/K-factor), in dm3.

αm – Flow meter volumetric thermal expansion coefficient, in ºC-1.

TLm – Water average temperature in flow meter, in ºC.

TC – Calibration temperature on flow meter, inºC.

δFm – Random variation on determination of Fm

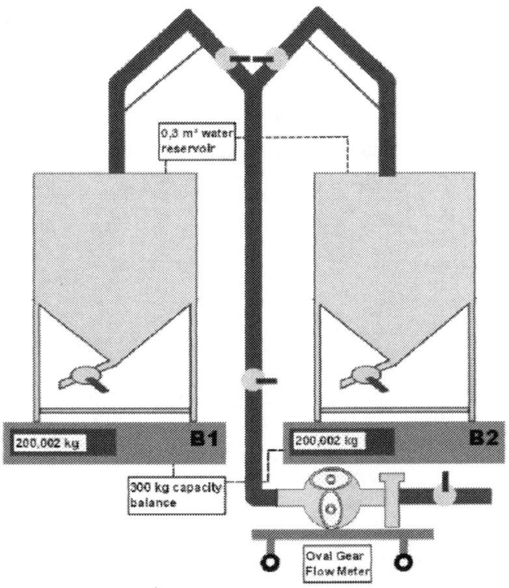


**Figure 1.** Weighing method calibration layout

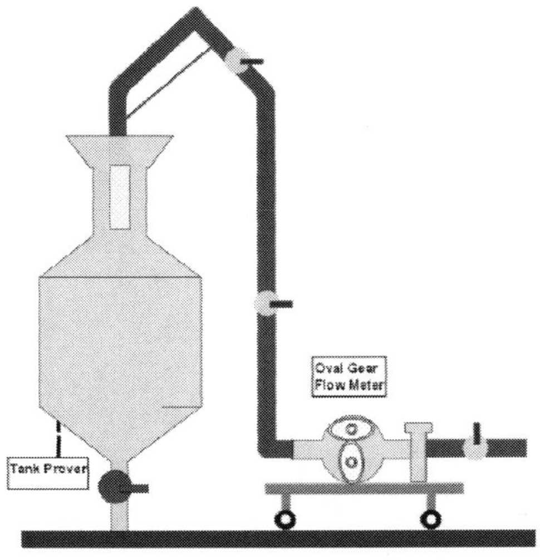


**Figure 2.** Volumetric method calibration layout

## 3. Results

The difference between the two methods was approximately 0.013%. The measurement uncertainty obtained in the weighing method was 0.019% while the uncertainty obtained by the volumetric method was 0.052%. In table 1 and 2, the contributions to the measurement uncertainty are shown individually, so observe the largest contributions to the uncertainty.

**Table 1.** Weighing method uncertainty budget

| Influence quantities | Relative contribution |
|---|---|
| Flow meter indication | 43% |
| Repeatability | 23% |
| Water mass | 14% |
| Non adjustment polynomial | 12% |
| Temperature variation during calibration. | 4% |
| Others | 3% |
| Total | 100% |

**Table 2.** Volumetric method uncertainty budget

| Influence quantities | Relative contribution |
|---|---|
| Tank indication correction | 47% |
| Meter factor repeatability | 17% |
| Temperature gradient in tank | 7% |
| Flow meter | 7% |
| Temperature variation in tank | 5% |
| Temperature variation in flow meter | 5% |
| Tank temperature | 4% |
| Temperature variation during calibration | 3% |

| | |
|---|---|
| Tank temperature | 3% |
| Tank resolution | 2% |
| Others | 1% |
| Total | 100% |

The calibration by the volumetric method was performed after the flow meter undergoing a course of about 100 m³, in a different rig which was used in the weighing method.

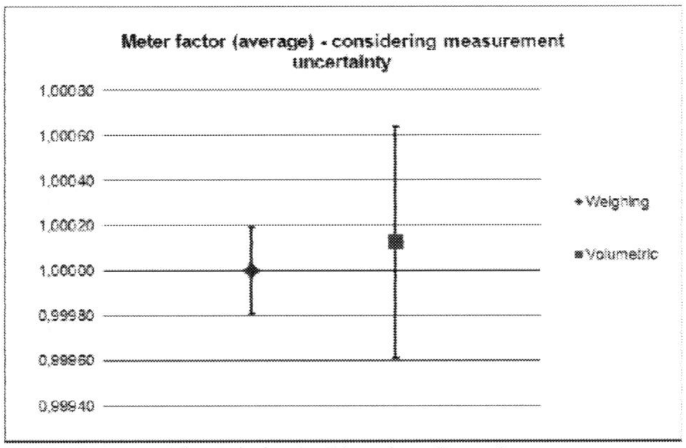


**Figure 3.** Calibration Results

## 4. Conclusions

The final calibration results were compatible. The normalized error was 0.23 showing that a flow meter can be calibrated in the field for a tank prover without compromising the final result. The weighing method system presented a good performance calibration even in the field, proving the effectiveness of the developed device.

The expanded uncertainty is directly related to the conditions of the flow meter installation. Proper installation, environment and liquid temperature and flow conditions directly influences in the final result of the calibration.

## 5. References

[1] Aibe V Y, Gouveia J G, Farias E C C, Silva E. , *Calibrations of tank prover with master flow meter and by weighing method using a mineral oil*, I Mechanical Metrology International Congress, Rio de Janeiro, 2008. **75**

[2] Aibe V Y, *Tank Prover Calibration in Locus* , 16 th International Flow Measurement Conference FLOMEKO, Paris, 2013. **281**

[3] Davis R S, *Equation for the determination of the density of moist air* (1981/91), Metrologia 29, 67 (1992). Giacomo P, *Equation for the determination of the density of moist air* (1981), Metrologia 18, 33 -1982.

# Dispersion analysis of accelerometer shock calibrations by comparison using a new exciter developed at Inmetro

**Giancarlo B Micheli, Gustavo P Ripper, Ronaldo S Dias, Cauê D Ferreira**

Vibration Laboratory, National Institute of Metrology, Quality and Technology – Inmetro, Duque de Caxias, RJ, Brazil

E-mail: gbmicheli@inmetro.gov.br

**Abstract**. The applicability of a new mechanical shock exciter developed at the Vibration Laboratory of Inmetro for primary calibrations was analysed for calibrations of accelerometers by the comparison method. Using two back-to-back reference transducers and a single-ended general purpose accelerometer as device under test (DUT), calibrations were performed using the exciter developed at Inmetro, and the results were compared with those obtained with a commercial shock exciter. Two shock amplitudes and two pulse durations were applied to evaluate the dispersion of the DUT shock sensitivity. This analysis shows that the new exciter provides appropriate repeatability for shock sensitivity measurements, indicating a powerful system for generation of shock pulses.

## 1. Introduction

Accelerometer calibrations under impulsive excitations are of interest in several areas where measurements of mechanical shocks are needed, such as automotive industry, mechanical test laboratories, safety industry and engineering design companies. For example, the Brazilian test laboratories accredited for drop-testing of automotive helmets need metrological traceability for their measuring transducers under similar shock conditions that are applied during actual tests.

The objective of a transducer shock calibration is to determine its shock sensitivity under an acceleration pulse with a specific shape. Gaussian, half-sine or half-sine squared are the pulse shapes recommended by international standards [1, 2]. The shock sensitivity is evaluated by measuring simultaneously the reference acceleration pulse and the electric output of the accelerometer under test (DUT). Using a standard accelerometer to obtain the reference acceleration pulse is the principle of the calibration by comparison. On primary shock calibrations of accelerometers, the reference acceleration pulse is measured by a laser interferometric method which includes signal processing techniques as differentiation data processing using digital low-pass filtering [3].

The Vibration Laboratory (Lavib) of the Brazilian National Institute of Metrology, Quality and Technology (Inmetro) is currently improving a developed primary shock calibration system [4, 5] composed by a shock exciter, a heterodyne laser interferometer and acquisition/processing subsystems.

The Calibration and Measurement Capability (CMC) on shock calibration by comparison was obtained and has been maintained by Lavib using a commercial exciter Endevco 2925 Pop Shock Calibrator. This CMC comprises charge and voltage shock sensitivity of an accelerometer submitted to half-sine shock pulse with amplitude from 200 m/s² to 100000 m/s² and durations from 0.1 ms to 3 ms, with a relative expanded uncertainty of 2%. Aiming to improve this uncertainty and considering that the commercial exciter is twenty years old, it was decided to evaluate the application of the new primary shock exciter also on comparison shock calibrations. The performance analysis of the Lavib-developed exciter for this kind of calibration is presented in this paper.

Calibration results of a shock transducer performed on the Lavib shock exciter and on the commercial exciter were compared. Both of these shock exciters use pneumatic actuators to generate the shock pulse. The same signal conditioning amplifiers and the same data acquisition system were used in the two experimental setups. The objective was to evaluate the repeatability of some parameters of the generated pulses and compare the results. It was expected that the new exciter could allow the reduction of the measurement uncertainties by diminishing the influence of some factors, such as transversal vibration, that increase the dispersion of the results. Another interesting convenience is to have the capability to perform both primary and comparison calibrations using one single system.

## 2. The Lavib-developed shock exciter

The Lavib-developed shock exciter, presented in figure 1, consists essentially of two 40 mm diameter airborne stainless steel cylinders (hammer and anvil), a pneumatic cylinder used as excitation system and stoppers with dampers to mitigate the energy and to limit the strokes.

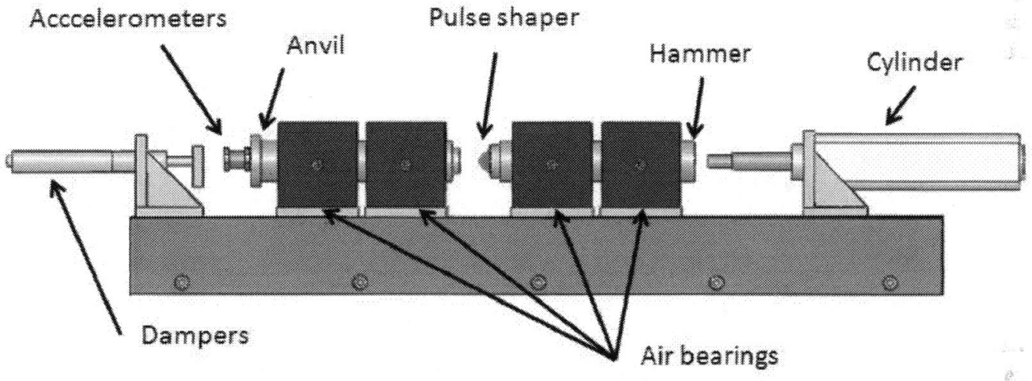


(a)

(b)

**Figure 1.** Scheme (a) and view (b) of Lavib shock exciter.

A pneumatic cylinder pushes the hammer which shocks against the anvil. A soft elastic interface (pulse shaper) is needed to shape the acceleration pulse properly. A back-to-back reference transducer can be fixed on an adaptor at the other end of the anvil. The single-ended accelerometer under test is mounted on the top reference surface of the back-to-back transducer. Some milliseconds after the pulse ends, the anvil motion is stopped by a damping system. Two hydraulic shock absorbers with adjustable damping rate are responsible for mitigating the kinetic energy of the anvil. The pulse amplitude can be controlled by the air pressure applied to the pneumatic cylinder and by the stiffness of the pulse shaper.

## 3. Experimental method

The experimental method used was to perform a calibration of a same transducer on the two exciter systems to compare the obtained results. The commercial shock calibrator (SE1) and the Lavib-developed exciter (SE2) were used to generate acceleration pulses with amplitudes of approximately 3500 m/s² and 500 m/s² and durations of 1 ms and 2 ms, respectively. All the measurement instruments used were the same for both systems, including the reference transducers, signal conditioning amplifiers, cables and data digitizing system.

Two Endevco 2270 back-to-back transducers, identified here as BTB1 and BTB2, were used as references to calibrate a single-ended accelerometer Endevco 2225 (DUT) for the 3500 m/s² tests. To evaluate the effect of smaller amplitudes, the BTB2 was applied for the 500 m/s² case.

The same data processing was applied to all signals acquired. The shock sensitivity of the accelerometer under test was calculated by taking the ratio of the charge amplitude generated by the accelerometer to the peak acceleration amplitude to which it was submitted. These amplitudes were calculated by the difference between the peak value and mean of the signal offset levels measured before and after the pulse. It was applied a $5^{th}$ order polynomial fit in the peak region of the pulse to identify a maximum value regardless the digital restriction of amplitude and time resolutions.

The pulse durations were calculated by the time difference in milliseconds at 10% of the peak amplitude of each pulse. Second order polynomials were fitted to the data in the ascending and descending regions of the pulse to evaluate the corresponding time values. Twenty measurements were taken for each case with a sample rate of 250000 Sa/s for 1 ms pulse durations and 100000 Sa/s for 2 ms pulse durations.

## 4. Measurement results

Typical pulses obtained using BTB2 as reference in the two exciter systems SE1 and SE2 are presented in figure 2. The curves relative to 3500 m/s² amplitude and 1 ms pulse duration are presented in figure 2 (a). The pulses relative to 500 m/s² amplitude and 2 ms duration can be observed in figure 2 (b).

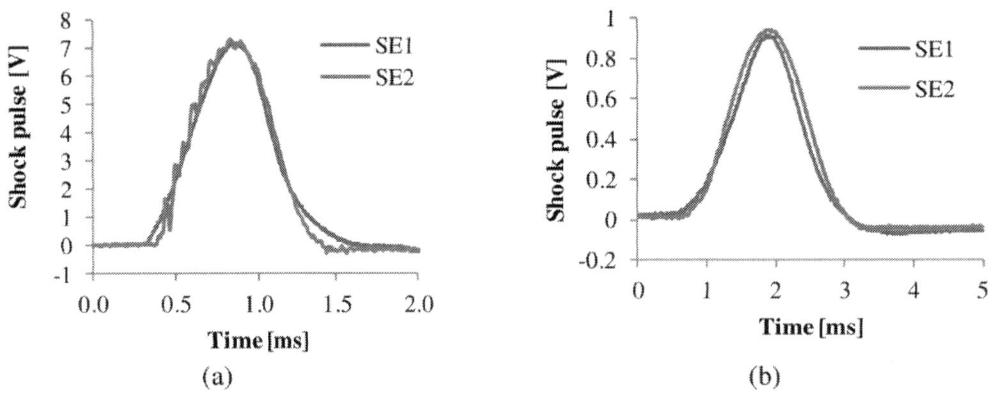


**Figure 2.** Shock pulses of (a) 3500 m/s² and (b) 500 m/s² amplitudes.

Mean values of the shock pulse parameters obtained from the data processing of the measurements, are presented in table 1 for the reference transducer BTB1 and in table 2 for BTB2.

**Table 1.** Calibration results for BTB1 as reference transducer.

| | 3500 m/s² (1 ms) | | |
|---|---|---|---|
| Quantity | SE1 | SE2 | Difference |
| Acceleration amplitude [m/s²] | 3578 | 3603 | + 0.7 % |
| DUT shock sensitivity [pC/(m/s²)] | 0.0884 | 0.0875 | - 1.00 % |
| Reference pulse duration [ms] | 0.97 | 0.85 | - 0.12 |
| DUT pulse duration [ms] | 0.98 | 0.86 | - 0.12 |

**Table 2.** Calibration results for BTB2 as reference transducer.

| | 3500 m/s² (1 ms) | | | 500 m/s² (2 ms) | | |
|---|---|---|---|---|---|---|
| Quantity | SE1 | SE2 | Difference | SE1 | SE2 | Difference |
| Acceleration amplitude [m/s²] | 3403 | 3338 | - 1.9 % | 437 | 462 | - 11.7 % |
| DUT shock sensitivity [pC/(m/s²)] | 0.0884 | 0.0879 | - 0.59 % | 0.0896 | 0.0891 | - 0.55 % |
| Reference pulse duration [ms] | 0.99 | 0.87 | - 0.12 | 1.95 | 1.93 | - 0.02 |
| DUT pulse duration [ms] | 1.01 | 0.88 | - 0.13 | 1.98 | 1.95 | - 0.03 |

It can be observed that considering the same reference transducer, the acceleration levels obtained were close for both exciters indicating a similar condition of excitation. The acceleration amplitude difference was larger for 500 m/s². Considering the parameter pulse duration, the closest results occur at 500 m/s².

In the three measurement cases, the DUT's shock sensitivity obtained on the new exciter system SE2 was lower than on the commercial machine SE1. The difference was greater with the reference accelerometer BTB1.

To analyze the repeatability performance, the standard deviations of measurements were computed and are presented in table 3 for the reference transducer BTB1 and in table 4 for BTB2. Relative percentage values are used to facilitate direct comparisons.

**Table 3.** Relative standard deviations (%) for BTB1 as reference transducer.

| | 3500 m/s² (1 ms) | |
|---|---|---|
| Quantity | SE1 | SE2 |
| Acceleration amplitude | 2.33 | 1.54 |
| DUT Shock Sensitivity | 0.18 | 0.20 |
| Reference pulse duration | 0.70 | 0.37 |
| DUT pulse duration | 0.94 | 0.63 |

Observing the results presented in tables 3 and 4 it can be noted that almost all the measurement standard deviations of the parameters analyzed here have decreased with the use of the new exciter system. Just the DUT shock sensitivity using the reference accelerometer BTB1 at 3500 m/s² presented a negligible difference.

Journal of Physics: Conference Series **648** (2015) 012017 doi:10.1088/1742-6596/648/1/012017

**Table 4.** Relative standard deviations (%) for BTB2 as reference transducer.

| | 3500 m/s² | | 500 m/s² | |
|---|---|---|---|---|
| Quantity | SE1 | SE2 | SE1 | SE2 |
| Acceleration amplitude | 1.60 | 0.94 | 1.65 | 1.54 |
| DUT Shock Sensitivity | 0.12 | 0.08 | 0.16 | 0.08 |
| Reference pulse duration | 0.60 | 0.31 | 0.53 | 0.30 |
| DUT pulse duration | 0.50 | 0.31 | 0.68 | 0.33 |

## 5. Discussion

In general, the new Lavib shock exciter result presented an appropriate repeatability to provide calibrations of accelerometers by comparison with expanded uncertainties lower than 2 %. This occurs probably due the use of small clearance air bearings and a precise control of the hammer propelling system. Air bearings maintain minimal friction along the movement of both hammer and anvil enabling higher repeatability of hammer excitation used to generate acceleration pulses on the anvil.

The pulse durations are basically a characteristic of the exciter, depending mainly on the pulse shaper components used. In this work it was taken the precaution to design pulse shapers to obtain similar pulse durations on the two exciters. This was important for a more reliable comparison between them.

The obtained differences between the two exciters on shock sensitivity, pulse amplitude and duration were considered satisfactory for the purpose of this repeatability analysis.

The ringing effect observed in figure 2 (a) for the SE2 curve at 3500 m/s² is due to excitation of an anvil longitudinal vibration mode by the shock. Initially, a low-pass filter had been applied to these signals to eliminate this effect. However, as the data processing already includes the polynomial fitting in the time domain, the results did not change significantly. Thus it was preferred to avoid the use of the low-pass filter.

The standard accelerometer BTB2 is more stable than the BTB1 and this could be observed on the results. The differences between the two exciters were less prominent when the reference BTB2 was used. This happened with the differences on shock sensitivity and on relative standard deviations of the acceleration pulse amplitude, shock sensitivity and pulse durations. The reason for this can be attributed to a flatter sensitivity response of the reference transducer BTB2 within the frequency ranges excited by the different pulse component frequencies, which were generated by the two shock exciters.

The flatness of the frequency-response sensitivity curve can also explain the difference on DUT shock sensitivity values between the two systems using the same reference transducer. Figure 3 presents the DUT sensitivity results obtained by sinusoidal excitation. The values are normalized by the sensitivity at 500 Hz. A half-sine acceleration pulse excites a broad range of frequencies in the accelerometer which depends on the actual pulse duration. Since the pulse durations obtained in each exciter was not exactly the same, the range of frequencies excited in each case was slightly different. Therefore, if the frequency-response sensitivity is not flat, the shock sensitivity may have been affected, explaining the differences between the sensitivity results of the two exciters.

Another feasible reason for the differences found on shock sensitivity values could be the unknown influence of axial vibration of the developed exciter anvil on the response of the transducers. The reliability of this phenomenon influence was diminished with the results obtained for 500 m/s² peak amplitudes. It was not observed ringing on the 500 m/s² amplitude pulse but it presented the same difference in shock sensitivity from the commercial exciter results.

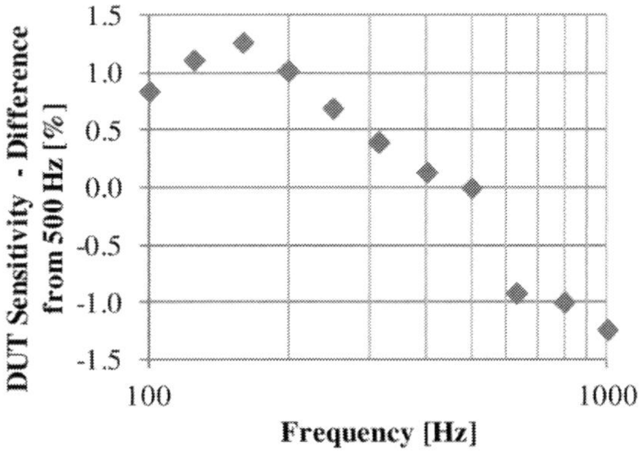


**Figure 3.** Differences of DUT sinusoidal sensitivity from the value at 500 Hz.

## 6. Conclusion

This work has presented the performance of a new mechanical shock exciter developed at the Vibration Laboratory of Inmetro for accelerometer calibrations by comparison. The study was developed making calibrations of a single-ended transducer in the new system and in a commercial machine that has been used by Lavib for customer's transducer shock calibration. The performance of these two exciters was compared by evaluating the repeatability of acceleration amplitude and duration of the pulses and also of the shock sensitivity values obtained them. The same auxiliary instrumentation as for instance signal conditioning amplifiers, cables and digitizing devices were employed to acquire the generated pulses in both systems.

Calibrations were carried out using two different back-to-back reference accelerometers. The results obtained with the new shock exciter have indicated that the developed apparatus presents an improvement in repeatability to generate shock pulses. Using the reference transducer BTB1, the relative standard deviations of the acceleration pulse amplitude, shock sensitivity, reference and DUT pulse durations were, respectively, 1.5 %, 0.2 %, 0.4 % and 0.6 %. For the reference transducer BTB2, two sets of calibrations were carried out. At 3500 m/s², the pulse amplitude presented a relative standard deviation less than 1% and for 500 m/s² the value was 1.5 %. The relative standard deviations of shock sensitivity and of reference and DUT pulse durations were the same values for the two amplitudes: 0.08 %, 0.3% and 0.3 %, respectively.

This study demonstrates that the new system despite of being developed mainly for primary shock calibrations can also be used for comparison calibrations. Further measurements will be carried out using a flatter sensitivity DUT to investigate the shock sensitivity differences between the systems, excluding this component of influence.

## References

[1] ISO 16063-13: 2001 Genève: International Standardization Organization
[2] ISO 16063-22: 2005 Genève: International Standardization Organization
[3] Nozato H, Bruns T, Volkers H and Oota A 2013 A study of Savitzky-Golay filters for derivatives in primary shock calibration *Acta Imeko* **2** 2 pp 41-47
[4] Micheli G B, Máximo L A and Ripper G P 2013 Um estudo sobre a influência da geometria e do material usado em uma máquina de choque mecânico para geração de pulsos de aceleração para calibração de acelerômetros *Proc. SAE Brazil Congress 2013* (São Paulo)
[5] Micheli G B, Garcia G A and Ripper G P 2010 Implementation of a Primary Shock Calibration System in Lavib - Inmetro *Proc. Simposio de Metrología* (Santiago de Querétaro) pp 27-9

# A proposal for dynamic calibration of brake tester

**Paulo L S Ferreira[1], Paulo R G Couto[1]; Luiz C Cabral[1], Ronaldo G Reis[1], Marcos Zillner [2]**

[1] Instituto Nacional de Metrologia, Qualidade e Tecnologia (Inmetro) Rio de Janeiro, Brasil
[2] Napro Eletrônica Industrial LTDA, São Paulo, Brasil

E-mail: plferreira@inmetro.gov.br [1]

**Abstract.** In Brazil there are about 400 security inspection lines carrier operating in Inspection Bodies accredited by Cgcre Inmetro [1]. The equipment in this proposal is a Brake Tester that measure vehicle braking forces and it is a component of an inspection line. This paper proposes a dynamic Brake Tester calibration using a reference torque transducer. This article can also be the basis for future discussions of the revised standard manufacturing of vehicle inspection line according to ABNT NBR 14040 [2].

## 1. Introduction

The Brazilian standard that specifies the manufacture of vehicle inspection lines (Figure 1) ABNT NBR 14040 [2] does not include metrological criteria to be met and also does not mention how the equipment must be calibrated.

```
1;2 Side-slip test
3 Shock absorber
inspection
4 Brake tester
5 Lift unit
6 Gas analyser
7 Opacimeter
8 Light alignment
```

**Figure1- Example vehicle inspection line**

The traditional method of Brake Testers calibration used by manufacturers is static. Consists of applying a calibrated weight placed on the end of the arm installed on the electric motor traction shaft Brake Tester these methodology is not efficient for calibration because it does not consider the actual dynamic vehicle in operation and the variables corresponding to the roller mechanism of the Brake Tester, as well as friction characteristics of the tires and brake friction elements, etc.

## 2. Purpose

This work aims to present a dynamic methodology calibration of a Brake Tester, figure 2. In this methodology the Brake Tester is calibrated by comparing its indicated strength with a reference standard torque transducer and the use of a special vehicle to meet the conditions of the test.

The methodology can serve as a basis for establishing metrological tolerances for Brake Testers, considering the linear and angular coefficients of the calibration curves of these instruments and the measurement uncertainty.

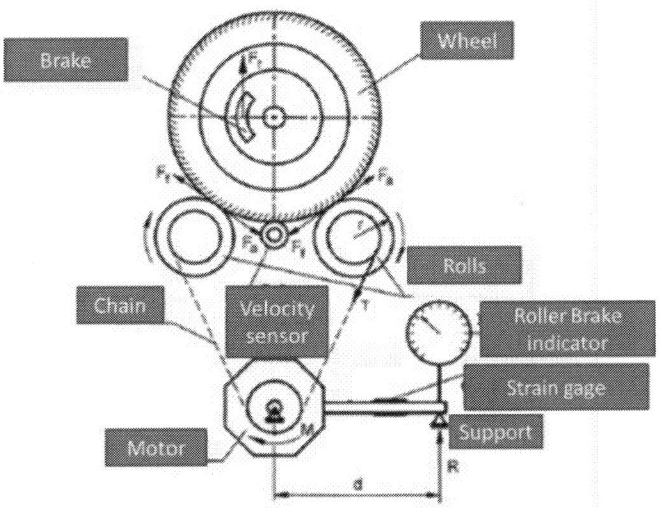


**Figure.2- Brake Tester**

## 3. Methodology proposal

*3.1. Performance of the test*

The performance of the test as proposed it consists the installation of a calibrated torque transducer and an encoder on the vehicle wheel axle for measurements of force and its tangential speed.
Another Brake Tester encoder is placed on the traction roll to measure its tangential velocity, as shown in Figure 3. This set will be called brake reference.

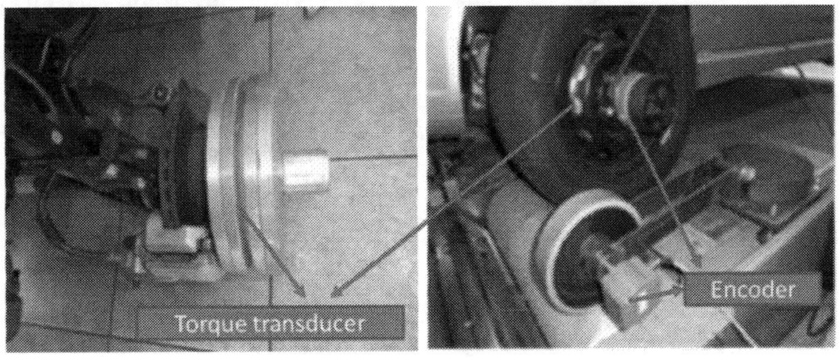


**Figure.3- Elements of brake reference**

In brake reference the wheel and encoders measure the different roller speeds, however before the start of braking there is no slippage between the roller and the wheel. Thus, the respective circumferential speeds are equal. Thus, this equation can be written by equation 1.

$$\omega_{roll} \cdot \pi \phi_{roll} = \omega_{wheel} \cdot \pi \phi_{wheel} \tag{1}$$

Where:

$\omega_{roll}$ = The roll angular velocity (rpm);

$\omega_{wheel}$ = Wheel angular velocity (rpm);

$\phi_{roll}$ = Roller diameter (mm);

$\omega_{wheel}$ = Wheel Diameter (mm);

Of equation above are known: the roll speed, roll diameter and wheel speed. Thus, the effective diameter of the wheel is defined by the equation 2.

$$\overline{\phi_{wheel}} = \frac{\omega_{roll} \cdot \phi_{roll}}{\omega_{wheel}} \tag{2}$$

The wheel diameter should be calculated according to the equation 2, because there is a deformation of the tire due to its support on the roll. Known effective diameter of the wheel and therefore its radius, the reference torque is converted into power which reference is compared with the strength indication Brake Tester.

The purpose of Brake Tester as control and monitoring tool is to record the maximum braking force that is obtained when the wheel speed reaches 20% to 30% of the roller speed [2]. At full strength, the measurement process is quite turbulent and this instability prevents one Brake Tester repeatability study this measuring point. In order to avoid instability, measurements are carried out in stages. A threshold means to maintain a constant nominal power brake by a predetermined period of time (Figure 4) called baseline, which can perform the registration of the force measured by Brake Tester compared with the reference value measured by the torque transducer.

At the beginning of the measurements indicates Brake Tester forces before the brake application by the driver, which forces are called parasites. These forces correspond to the resistance to movement system (rollers, wheels, transmission system, etc.) that is registered by the load cell Brake tester and are not detected by Brake reference. The parasitic forces should be discounted in the mathematical model that defines the amount recorded by Brake Tester when compared to the value as equation 3.

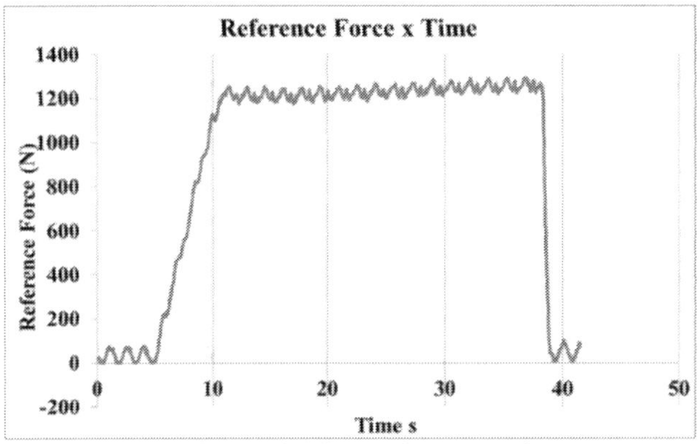


**Figure.4-Measuring Baseline**

For the test to be carried out under reproducibility conditions of the test operators (vehicle driver and technician responsible for measurements) were previously trained. Also a special vehicle to meet the conditions of the test with brake system working properly. The vehicle tire pressure were set at 45 psi (310 kPa) before each test. During the test run, the vehicle was moved from one installation to another towed. The measurement was set out of the rear axle, which is rigid with measurements results in lower dispersion values. The reference standard installed on the axis of the vehicle was assembled and disassembled each experiment by trained technician to avoid potential problems on shipping. The braking of the vehicle during each test was performed by the operator of the evaluated inspection line.

### *3.2 Establishment of the calibration curve model*

Whereas the vehiclè used produces a maximum braking force on the order of 1800 N, was chosen four set points force (measuring levels) in 400N; 800N; 1100N and 1400N. The test was performed in descending order of strength. For each nominal point of measurement were considered five medium arising from an acquisition of 20 Hz rate (20 samples per second) continuous measurement that equals approximately 200 values.

In possession of all the results of averages for each of the nominal values and their reference values, Brake Tester calibration curve can be defined by the model of equation 3:

$$F_{ref} = a + b \cdot \left(F_i - F_p\right) \tag{3}$$

Where:

$a$ = linear coefficient (C);
$b$ = slope;
$F_i$= Brake Tester force indicated by (N);
$F_p$= Force parasite (N).

The systematic error of Brake Tester was evaluated by linear coefficient of its calibration curve according to the new methodology proposed by this work. The slope measures the specific property of each Brake Tester, comprising the drive mechanism of the rollers, placement of meters and sensors as geometry of each project, etc.

### *3.3 Validation of the calibration curve model*

The validation of the Brake Tester calibration curve model was carried out by residual analysis and variances [3].

### *3.4 Measurement uncertainty*

Estimation of the calibration of a Brake Tester measurement uncertainty followed the ISO GUM recommendations in 2008 [4].

Journal of Physics: Conference Series **648** (2015) 012018 doi:10.1088/1742-6596/648/1/012018

*3.4.1 Measuring*

The measured calibration of an Brake Tester was defined by the force value that evaluated the timely systematic error of the instrument, according to equation 4.

$$\overline{e_i} = \overline{F_i} - \overline{F_p} - \overline{F_{ref}} = \overline{F_i} - \overline{F_p} - \left[\frac{\overline{\Delta V} - a_1}{b_1}\right] \cdot \frac{2}{\overline{\phi_{wheel}}} \qquad (4)$$

Where:

$\overline{F_i} =$ Medium strength indicated by Brake Tester (N);

$\overline{F_p} =$ Force average parasite (N);

$\overline{F_{ref}} =$ Reference average force (N);

$a_1 =$ Linear coefficient of the calibration curve of the torque transducer (V);

$b_1 =$ Slope of the torque transducer calibration curve (V / Nm);

$\overline{\Delta V} =$ Medium voltage indicated by the transducer (V) calibration;

$\overline{\phi_{wheel}} =$ Mean wheel diameter (m) according to equation 2;

*3.4.2 Cause-Effect Diagram*

According to equation 4, the uncertainty sources are shown in the diagram cause effect of Figure 5:

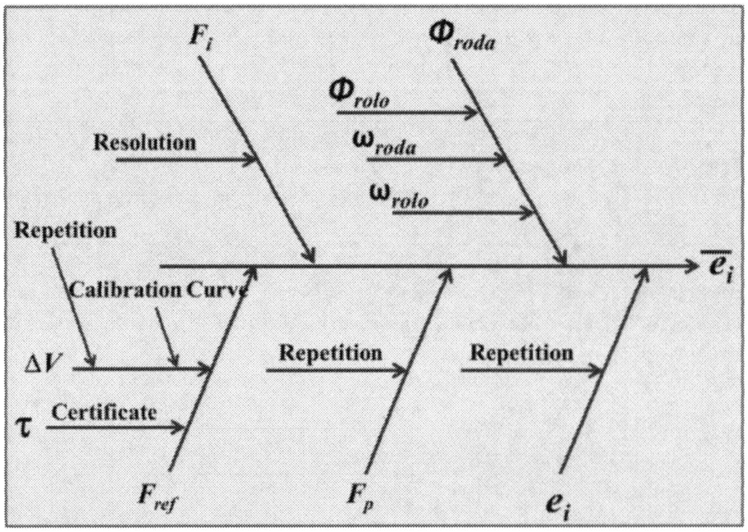


**Figure. 5 Diagram Cause Effect of estimated Brake Tester error uncertainty**

*3.4.3. Uncertainties of the input quantities*

*3.4.3.1 Wheel diameter*

The uncertainties related to roll angular velocity (ω roll) of the wheel (ωwheel) and roll diameter are estimated using the equations 5, 6 and 7, respectively.

$$u_{\varpi\, roll} = \left( \sum_{i=1}^{5} \left( \frac{s_{\omega i\, roll}}{n_{\varpi i\, roll}} \right)^2 + \frac{a^2_{resolution\, \varpi\, roll}}{3} \right)^{1/2} \quad (5)$$

$$u_{\varpi\, roda} = \left( \sum_{j=1}^{5} \left( \frac{s_{\varpi i\, wheel}}{n_{\varpi i\, wheel}} \right)^2 + \frac{a^2_{resolution\, \varpi\, wheel}}{3} \right)^{1/2} \quad (6)$$

$$u_{\phi\, roll} = \frac{a_{resolution\, Droll}}{\sqrt{3}} \quad (7)$$

Where:

$s_{\omega i_{roll}}$ = Standard deviation of the angular roll speed values;

$n_{\omega i_{roll}}$ = Number of repetitions of the values of the angular speed of the roll;

$a_{resolution\, \omega_{roll}}$ = Scanning resolution of the angular speed of the roll;

$s_{\omega i_{wheel}}$ = Standard deviation of the angular wheel speed values;

$n_{\omega i_{wheel}}$ = Number of repetitions of the angular wheel speed values;

$a_{resolution\, \omega_{wheel}}$ = Scanning resolution of the angular speed of the wheel;

$a_{resolution\phi roll}$ = Reading resolution of the roll diameter.

From equation 2 to the uncertainty of the wheel diameter is then defined by Equation 8.

$$u_{D_{wheel}} = \left( \left( \frac{\partial \phi_{wheel}}{\partial \omega_{roll}} \cdot u\omega_{roll} \right)^2 + \left( \frac{\partial \phi_{wheel}}{\partial \phi_{roll}} \cdot u\phi_{roll} \right)^2 + + \left( \frac{\partial \phi_{roda}}{\partial \omega_{wheel}} \cdot u\omega_{wheel} \right)^2 \right)^{1/2} \quad (8)$$

Journal of Physics: Conference Series **648** (2015) 012018 doi:10.1088/1742-6596/648/1/012018

*3.4.3.2 Strength Indicated by Brake Tester Fi*

*3.4.3.3 Uncertainty on the settlement*
The uncertainty regarding the resolution of force indicated by Brake Tester is estimated by equation 9.

$$u_{Fi} = \left( \frac{a^2_{resolution\ Fi}}{3} \right)^{1/2} \tag{9}$$

Where:

$a_{resolution\ Fi}$ =Scanning resolution of the indicated strength.

*3.4.3.4 Force Parasite Fp.*

*3.4.3.4.1 Uncertainty on the settlement*
The uncertainty related to the parasite replicates force is estimated by equation 11.

$$u_{Fp} = \left( \frac{s^2_{repetition\ Fp}}{n} \right)^{1/2} \tag{10}$$

Where:

$s_{repetition\ Fp}$ = Standard deviation of the values of parasite force;

$n$ = Number of repetitions of the values of parasite force for each nominal value

.

*3.4.3.5 Torque transducer $\tau_{ref}$*

*3.4.3.5.1 Uncertainty regarding the calibration certificate of the torque transducer*
The uncertainty regarding the calibration certificate of the torque transducer is definede by the equation (11)

$$u_{certificate}\tau = \frac{U}{100 \cdot k} \cdot \tau_{ref} \tag{11}$$

Where:

$U$ = Transducer Certificate uncertainty (%);
$k$ = Certificate coverage factor;

*3.4.3.6 Uncertainty regarding the measurement of voltages*

Uncertainty regarding the power indicated by the torque meter is estimated by equation 12.

$$u_{\Delta V} = \left( \sum_{i=1}^{5} \left( \frac{s_{\Delta Vi}}{n_{\Delta Vi}} \right)^2 + \frac{a^2_{resolution\,\Delta V}}{3} + u^2_{\Delta Interpolation} \right)^{1/2} \tag{12}$$

Where:

$s_{\Delta Vi} =$ Standard deviation of the voltage values indicated by the transducer for each set;

$n_{\Delta Vi} =$ Number of repetitions of the voltage values indicated for each set;

$a_{resolution\ \ \Delta V} =$ Voltage reading resolution

According to [2] the uncertainty related to transducer calibration curve is estimated by equation 13.

$$u_{\Delta\ Interpolation} = S_e \cdot \left( \frac{1}{m} + \frac{1}{n_t} + \frac{\left(\tau_o - \overline{\tau_i}\right)^2}{\sum\left(\tau_i - \overline{\tau_i}\right)^2} \right)^{1/2} \tag{13}$$

Where:

$S_e =$ Transducer curve of dispersion (V);

$m =$ Number of repetitions of the measurement voltage measurements;

$n_t =$ Number of points transducer certificate;

$\tau_o =$ Average torque values of the calibrated point;

$\tau_i =$ Torque values indicated in the calibration certificate;

$\overline{\tau_i} =$ Average voltage values specified in the certificate.

*3.4.3.7. Uncertainty relating to the repetitions of machine errors*

The uncertainty concerning the dispersion of Brake tester errors is estimated by equation 14

$$u_{ei} = \frac{s}{\sqrt{n}} \tag{14}$$

Where:

$s =$ Standard deviation of Brake Tester errors;

$n =$ Number of repetitions

Journal of Physics: Conference Series **648** (2015) 012018 doi:10.1088/1742-6596/648/1/012018

*3.4.4 Sensitivity Coefficients*

From equation 4, the measuring sensitivity coefficients are calculated by the equations 15, 16, 17,18 and 19.

$$c_i(F_i) = \left(\frac{\partial \bar{e}_i}{\partial F_i}\right) = 1 = c_1 \tag{15}$$

$$c_i(Certificate\ \tau) = \left(\frac{\partial \bar{e}i}{\partial Certificate\ \tau}\right) = -\frac{2}{\phi wheel} = c_2 \tag{16}$$

$$c_i(\overline{\Delta V}) = \left(\frac{\partial \bar{e}i}{\partial \overline{\Delta V}}\right) = -\frac{2}{b_1 \cdot \phi_{wheel}} = c_3 \tag{17}$$

$$c_i(\phi_{wheel}) = \left(\frac{\partial \bar{e}i}{\partial \phi_{wheel}}\right) = \frac{2\left(\overline{\Delta V} - a_1\right)}{b_1 \cdot \phi_{wheel}^2} = c_4 \tag{18}$$

$$c_i(e) = \left(\frac{\partial \bar{e}_i}{\partial e_i}\right) = 1 = c_5 \tag{19}$$

*3.4.5 Uncertainty combined calibration of Brake Tester*

The combined uncertainty of the systematic Brake Tester error is defined by the equation 20.

$$u_c(\bar{e}_i) = \left( \left(c_1 \cdot u_{Fi}\right)^2 + \left(c_2 \cdot u_{Certficate}\tau\right)^2 + \left(c_3 \cdot u_{\overline{\Delta V}}\right)^2 + \left(c_4 \cdot u_{\phi wheel}\right) + \left(c_5 \cdot u_{ei}\right)^{2^2} \right)^{1/2} \tag{20}$$

*3.4.6 Number of effective degrees of freedom*

The number of effective degrees of freedom is defined by equation 21.

$$v_{eff} = \frac{\left[u_c(\bar{e}_i)\right]^4}{\sum_{j=1}^{5}\left(\frac{c_1 \cdot u_{\Delta Vi}}{v_{\Delta Vi}}\right)^4 + \left(\frac{c_4 \cdot u_{\phi r}}{v_{\phi wheel}}\right)^4 + + \left(\frac{c_5 \cdot u_{ei}}{v_{ei}}\right)} \tag{21}$$

Where:

$u_{\Delta Vi} =$ Uncertainties relating to each set of repetitions indicated voltage;

$v_{\Delta Vi} =$ Number of degrees of freedom for each set of replicates and the voltage curve;

$v_{\phi wheel} =$ Number of degrees of freedom of the wheel on the combined uncertainty.

*3.4.7 Coverage Factor*

The coverage factor (k) is defined as the number of effective degrees of freedom computed by the equation 19 and a probability of 95% coverage.

## 4. Results

The methodology described for the estimation of uncertainty of measurement was applied to calibrate an Brake Tester only a nominal point of 1400 N according to Equation 4 in Table 1 shows the values of the quantities input defining the measurement uncertainty of Brake Tester strength to the point 1400 N.

**Table 1.** Input quantities defining the calculation of the force measured by a Brake Tester

| *Sources* | *Value* | *Unit* |
|---|---|---|
| ***u Fi resolution*** | 0,577 | N |
| ***u error*** | 10,58 | N |
| ***uFp*** | 0,000 | N |
| ***u*** $\tau$ ***certificate*** | 0,000 | Nm |
| ***u*** $\Delta V$ | 0,001 | V |
| ***u*** $\Phi$ ***wheel*** | 0,002 | m |

Uncertainty components for each input variable and the combined strength uncertainty measured by Brake Tester are presented in Table 2.

**Table 2.** Uncertainty components and combined uncertainty.

| Components | *N* | % |
|---|---|---|
| ***uF***(Fi) | 0,6 | 0,2 |
| ***uF(Fp)*** | 0,0 | 0,0 |
| ***uF(erros)*** | 10,6 | 71,0 |
| ***uF(certificado)*** | 0,0 | 0,0 |
| ***uF***($\Delta V$) | 5,3 | 17,7 |
| ***uF(froda)*** | 4,2 | 11,1 |
| *uc* | **12,6** | |

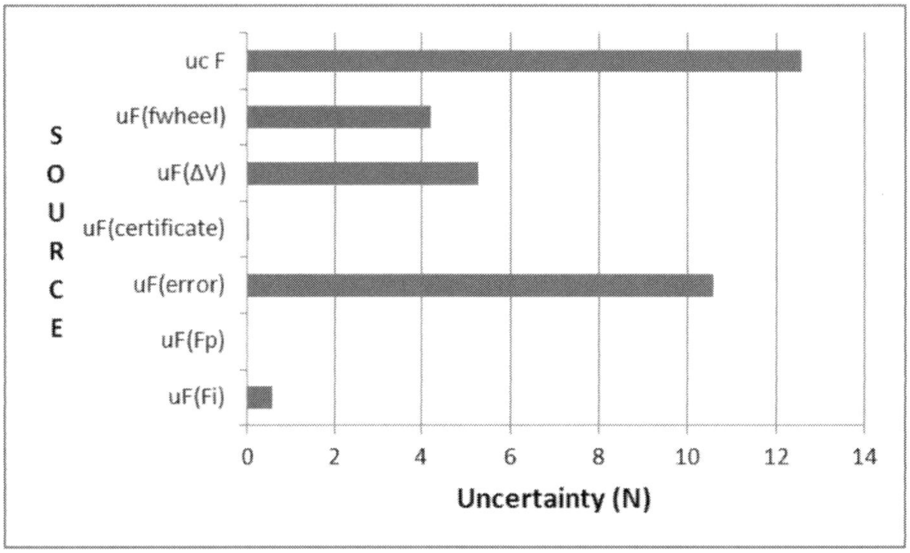


**Figure. 6 - Force uncertainties balance measured by Brake Tester**

Table 3 shows the components of uncertainty in the measured voltage with their respective contributions.

**Table 3.** Uncertainty contributions in Volts

| *Source* | *Uncertainty(V)* | *Contribution (%)* |
|---|---|---|
| $u\ \Delta V1$ | **3,2 x 10$^{-04}$** | **8,2** |
| $u\ \Delta V2$ | **3,1 x 10$^{-04}$** | **7,4** |
| $u\ \Delta V3$ | **2,7 x 10$^{-04}$** | **5,7** |
| $u\ \Delta V4$ | **2,9 x 10$^{-04}$** | **6,9** |
| $u\ \Delta V5$ | **2,6 x 10$^{-04}$** | **5,3** |
| $u_a$ ***resolution*** $\Delta V$ | **5,8 x 10$^{-05}$** | **0,3** |
| $u\ \Delta$ ***Interpolation*** | **9,1 x 10$^{-04}$** | **66,5** |
| $u\ \Delta V$ | **1,1 x 10$^{-03}$** | |

Applying the values in equation 4 of force indicated by Brake Tester parasite strength of the electric voltage measured at the torque transducer and wheel diameter to the nominal point of 1400 N obtains the measurement result measured by the force Brake Tester

$$1085\ N \pm 14N\ (p=95\%;\ k=2{,}03)$$

In uncertainty (uc F) should be added to the Brake Tester the contribution regarding its calibration curve when it is used as a reference for calibrating the other line inspection. The uncertainty concerning the reference Brake Tester calibration curve is calculated by the expression 22.

$$u_{curve\,Fi} = S_{eF} \cdot \left( \frac{1}{m_F} + \frac{1}{n_{tF}} + \frac{\left(F_o - \overline{F_i}\right)^2}{\sum\left(F_i - \overline{F_i}\right)^2} \right)^{1/2} \tag{22}$$

Where:

$S_{eF}$ = Dispersal Brake Tester curve reference (N);
$m_F$ = Number of repetitions of the force measurements indicated by reference Brake Tester;
$n_{tF}$ = Number of points of reference Brake Tester certificate;
$F_o$ = Average strength values indicated by the reference in the calibrated point;
$F_i$ = Strength values given in the calibration certificate in reference Brake Tester;
$\overline{F_i}$ = Average strength values given in reference Brake Tester certificate

## 5. Conclusions

The values presented show that the nominal point of 1400 N, the predominant element of uncertainty is that regarding the dispersion of machine defects, with a contribution of 71% of the combined force uncertainty.

According to Table 3, the uncertainty of interpolation which comes from the meter calibration certificate torque is the uncertainty that prevails in the electric voltage and whose contribution is approximately 66.3%.

The expanded the aforementioned measurement result uncertainty is approximately 2.7% of the force measured by Brake Tester.

From the obtained measurement result can be observed that the strength of 1400 N indicated by Brake Tester has an error of approximately 22%, which leads to the conclusion that in the current state of the art improvements must be made in equipment for the attenuation of this error, taking into account the uncertainty obtained.

The proposed methodology can be:

• Basis for the realization of inter comparisons in order to continuously be upholding metrological reliability of Brake Testers in the country;

• Used as a tool for the technical management of Brake Testers calibrations as well as aid in the technical guidance for revision of ISO 14040 standard.

Improving the work can be obtained from the acquisition and sync Brake Tester statement data continuously, with the possibility of lowering the instrument error.

## 6. Acknowledgements

Thanks to everyone who participated in the preparation of this work in INMETRO namely: Renata Martins Horta Borges; Guilherme Andrade Campos Pedrosa; Tiago Dantas de Oliveira and especially to Napro company, which enabled the construction and development of the standard and reference standard vehicle used in this study.

## 7. References

[1]*http://www.inmetro.gov.br/organismos/resultado_consulta.asp*

[2] *Norma ABNT NBR 14040, Inspeção de segurança veicular, Veículos leves e pesados.*

[3] *Estatística Aplicada e Probabilidade para Engenheiros- Montgomery D. C ; Runger G C – Segunda Edição – 2003*

[4] *Guia para a Expressão da Incerteza de Medição. ISO GUM 95. Terceira Edição Brasileira Guide to the Expression of Uncertainty in Measurement. Rio de Janeiro, ABNT, INMETRO, Rio de Janeiro, 2008.*

# Estimate of the inertial torque in rotating shafts - a metrological approach to signal processing

**R Oliveira[1,2] , R Claudino[1], R Kalid[1], T Fröhlich[3], L Gusmão[1], H Lepikson[1]**

[1]Universidade Federal da Bahia (UFBA), Salvador - BA, Brasil
[2]Instituto Nacional de Metrologia, Qualidade e Tecnologia (INMETRO), Rio de Janeiro - RJ, Brasil
[3]Technische Universität Ilmenau (TU-Ilmenau), Ilmenau, Alemanha

E-mail: rsoliveira@inmetro.gov.br

**Abstract.** This article studies the setting of the method to model the inertial torque in the acceleration regimes of a rotating axis system. The angular speed measurement to obtain an estimated acceleration and the moment of inertia axis of the variables are used to calculate the inertial torque. The evaluated dynamic regimes are acceleration ramps between levels (intervals) of angular speed, which are implemented using variable speed drives that are coupled to AC motors. The article focuses on assessing the main parameters in signal processing, such as the derivation of the speed signal, application of digital filters and data-smoothing methods, and their effects on the main features of interest in the generated torque curves, such as the peak torque amplitude, ascending and downward slopes, stability and repeatability. A qualitative approach was performed to identify the measurement uncertainty contributions. Thus, the validation methodology of the processing methods and the proposed physical principle are feasible.

## 1. Introduction

In systems or machines that use rotating shafts, the torque and angular velocity magnitudes are fundamentally important to the operation and characterization of most processes [1, 2, 3]. Examples of these operations are the positioning of a load, calculation of the transmitted mechanical power and restrictions on the acceleration regimes of equipment because of the inertial torque.

The load torque contributions can be estimated based on the direct correlation of weights, bending moments and assembly's friction characteristics for the displacements of load pieces. Therefore, its estimation is inserted to the definitions of the application's parameters such as the speed of operation and maximum efforts.

In the acceleration regimes, the inertial torque is an undesired behavior for setting the shaft driving. Thus, the angular acceleration regimes are often overdimensioned and function more as a safety and functional restriction when the acceleration periods are set to last long.

However, if the inertial torque response is more accurately estimated, the shaft driving can reach better performance with faster accelerations in different speed steps.

The angular velocity measurement with encoders is the usual method to estimate the inertial torque in the shaft instead of the direct measurement using torque or load sensors.

Journal of Physics: Conference Series **648** (2015) 012019 doi:10.1088/1742-6596/648/1/012019

There are two torques on the axis during acceleration: the so-called load torque, which is generated from the resistance from the load to the movement, and the inertial torque, which is generated from the angular acceleration of the shaft masses and those attached to it.

One method to estimate the inertial torque is to use the product of the angular acceleration and the moment of inertia of the shaft. Generally, the moments of inertia are estimated quantities, which can be obtained from technical specifications of the components that are attached to the shaft.

In turn, the angular acceleration is generally estimated based on the angular speed signal, which is measured from the shaft.When measuring time-varying mechanical quantities, a dynamic error is introduced, i.e., the ratio between the input and the response cannot maintain itself as in static measurements. To mitigate these errors, the measured signal is post-processed, and the system dynamics modeling can be introduced [4]. Thus, the choice of a method or parameter to process the signal is an important step because the same signal-processing process can obtain different answers for different portions of the curve, which will result in loss of information or conversely, an unnecessary computational burden.

This paper presents a comparative study of some measurement and data-processing methods for the speed and acceleration, which relate the results to the regions and points of interest in the estimated inertial torque curves.

The purpose of the article is not to set or select the best method but to quantitatively demonstrate the difference among the obtained results and present a methodology that can be reproduced to validate these proceedings when they are applied.

## 2. Methodology

The research methodology was based on the data acquisition and manipulation of speed and torque under certain dynamical regimes from driving a measurement shaft. The acceleration and torque profiles are similar to the load impulses, where the peaks are fast, and these quantities did not have large stabilization levels and rates of change. These characteristics determine the dynamic schemes.
According to the theory, the relationship between torque and angular acceleration should follow the physical principle of Newton's second law for angular movements, as shown in equation 1.

$$\tau = \theta \cdot \dot{\omega} \therefore \theta = \frac{\tau}{\dot{\omega}} \tag{1}$$

The experiments were performed at the Engine Test Laboratory of the Research Centre for Electric Energy (CEPEL). The bench (figure 1) consists of an electric motor, a variable speed drive, a torque transducer and an integrated speed sensor (incremental encoder), which were coupled in balance.

To define the type of regime under study, it is important to demonstrate the real profile of the curves or ramps controlled by the drives. The control parameters of the drives are adjusted for a particular application, so the inertial generated torque in the regime is used to calculate the minimum acceleration time. The theoretical equation obtained from the operating manuals of this equipment relates the rated power and rated speed with the acceleration and total moment of inertia to be accelerated [5]. However, this acceleration is assumed to be constant throughout the period, which is not consistent with the acceleration curves that are calculated from the angular-velocity measurements.

To illustrate this point, figure 2 shows two speed curves to fit the acceleration time of 2 s, which is set in the drive; one curve represents the theory, where the acceleration is considered constant, and the other curve represents the measured profile.

It is observed that the theoretical speed step is linear, which indicates a constant acceleration, which will approximately correspond to the mean of the measured acceleration, as shown in figure 3. This result is important because differences among the acceleration curves results in an even larger difference among the estimated inertial torques, which can determine the extra load on the shaft.

**Figure 1.** Test bench used in the experiment.

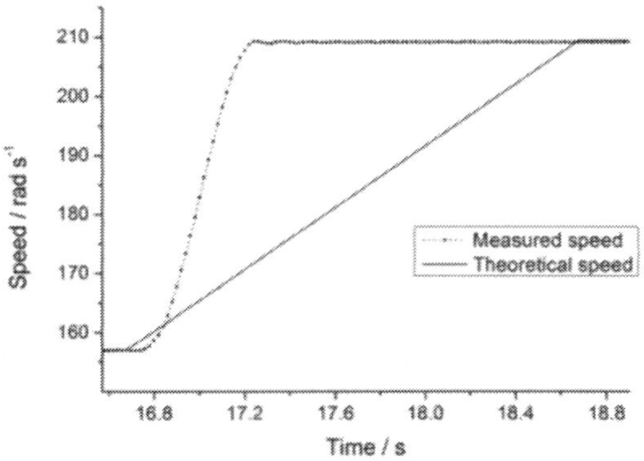


**Figure 2.** Acceleration ramp: measured speed and theoretical speed.

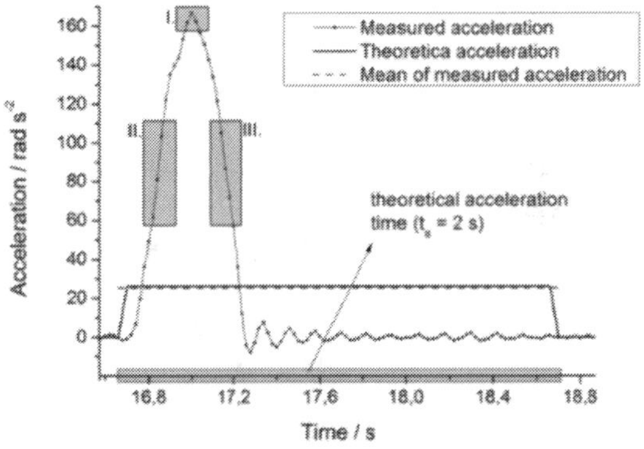


**Figure 3.** Acceleration curves: measured acceleration curve with identified critical regions and theoretical acceleration.

From the understanding of the applied real acceleration profile, the interesting regions of the curve can be determined to parameterize the inertial torque. In figure 3, these regions of interest were identified, considering the direct relationship between acceleration and torque: I) higher torque value (peak torque); II) higher positive torque rate; III) higher negative torque rate.

Then, the curves and methods were measured with the motor driving in steps of 52 rad $s^{-1}$ in three different speed intervals: 157-209 rad $s^{-1}$, 209-261 rad $s^{-1}$ and 261-313 rad $s^{-1}$.

The acceleration time was set to 2 s, and the driving curves were repeated three times in each speed interval.

From the measured speed data, different commonly used signal-processing methods and tools were applied. The results were statistically analyzed using their nominal and dispersion values, considering the features and regions of interest in the driving curves, such as the peak accelerations and intermediate regions in the ascending and descending slopes of the curve.

The data treatment methods that were used are the commonly performed signal-processing methods with the identical characteristic regime, such as the components of moderate intensity noise and non-periodic instantaneous phenomenon to be measured. Some methods were applied to the original speed data (raw data), whereas others were applied to the calculated acceleration signal, which was obtained from the derivation of the original speed signal. A brief description of each applied method [6] is shown below.

*Savitzky-Golay smoothing (SG)* - The method performs a local polynomial regression around each point and creates a new smoothed value for each data point. This method preserves the data characteristics such as peak height and width. The softness depends on the window size (number of points) to be used in the algorithm. For this analysis, windows with 15 points were used to smooth the original signals of the speed and the acceleration values. This method was also applied to smooth the torque signal.

*Boltzman* - This curve adjustment method produces a sigmoid response, which is notably similar to the driving results. It was only applied to the original speed signal.

*Gauss* – This method adjusts the data to a Gaussian curve, which is notably similar to the curves obtained for the acceleration profiles. It was only applied to the acceleration signals that are derived from the original speed signal.

*Polynomial fit* - Traditional method of polynomial regression. The software uses the least-square method. To make a comparison in the method, which is widely used in metrology, adjustments were made using a third-degree polynomial.

The measured torque values act as checker parameters of the adopted physical principle according to equation (1).

## 3. Results

The results are shown for each type of evaluation. Likewise, when appropriate, a brief conclusion or specific observations are shown. Subsections 3.1, 3.2, 3.3 and 3.5 are based on the same data that were measured on the three consecutive drives for each speed range at an acquisition rate of 50 Hz. For subsection 3.4, the acquisition rate varies as indicated.

### *3.1. Evaluation of the stability curves*

This subsection shows the evaluation of the measurement stability during the acceleration in different speed ranges. The standard deviation was calculated from the speed and acceleration curves that were adjusted to the same time interval, i.e., the beginning of each acceleration ramp corresponds to time t = 0 s of the acceleration period. Figure 4 shows four speed deviations, whereas figure 5 shows the calculated acceleration deviation.

It is observed that the speed evaluation obtains smaller dispersions in the entire period of the curve than the acceleration evaluation. However, the center of the acceleration curve, where the peak is located, has higher stability, which is exactly the opposite of the observed behavior of speed curves.

It is also observed that the acceleration data that were calculated from the raw speed data have higher variation in standard deviation than those obtained from the smoothed data, filtered and fitted curves, which confirms the motivation for this study. Another important observation is that both curves were more stable in the range of 209-261 rad $s^{-1}$.

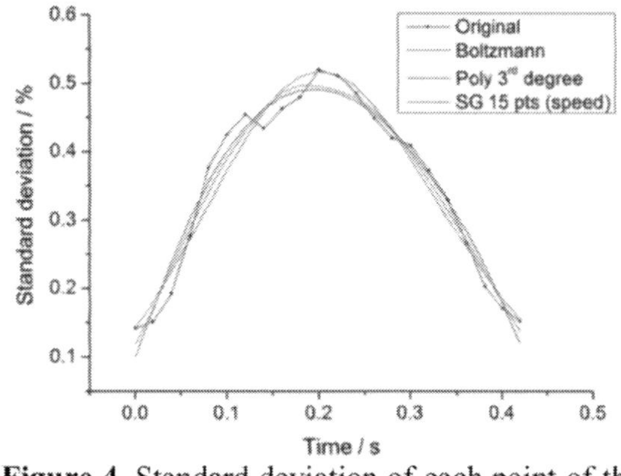


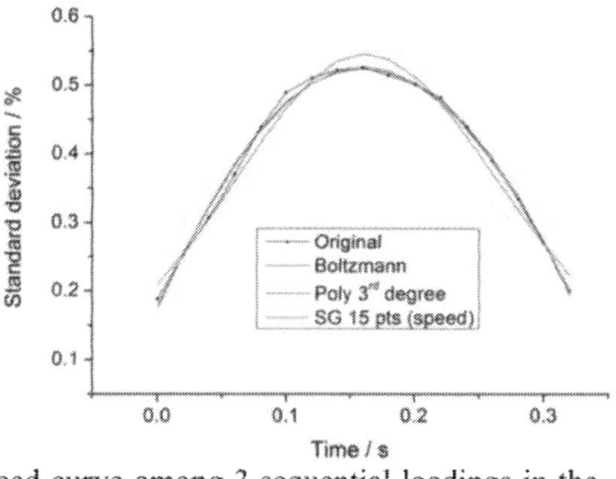


**Figure 4.** Standard deviation of each point of the speed curve among 3 sequential loadings in the speed intervals of (left) 157-209 rad $s^{-1}$ and (right) 209-261 rad $s^{-1}$.

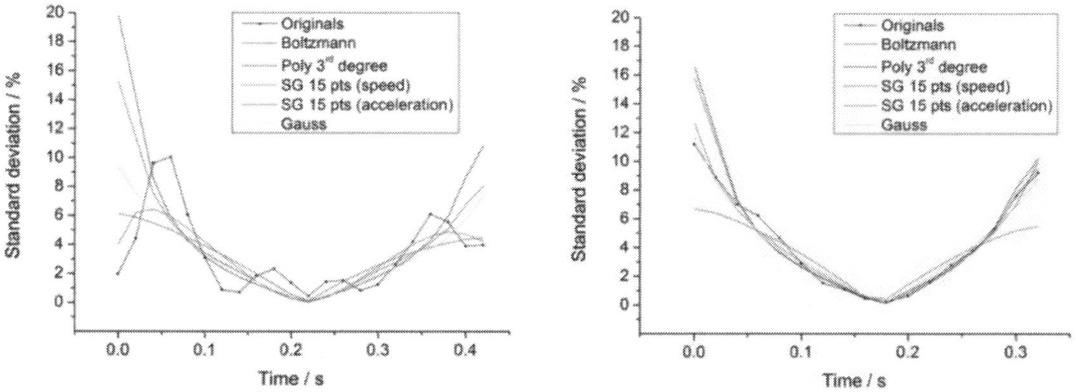


**Figure 5.** Standard deviation of each point of the acceleration curve among 3 sequential loadings in the speed intervals of (left) 157-209 rad $s^{-1}$ and (right) 209-261 rad $s^{-1}$.

### *3.2. Evaluation of peak acceleration*

The peak acceleration corresponds to the peak torque and is the most critical region of the curve in terms of mechanical stress of the rotating shaft. Its variability was evaluated in different speed ranges under different acquisition rates and the results of different signal-processing methods that were applied to the same mass of data.

#### *3.2.1. Peak variation among different speed ranges*

Theoretically, at the identical degree of speed, which is 52 rad $s^{-1}$ in this case, the drive should apply the identical curve regardless of the acceleration speed range. However, table 1 shows a difference of the calculated peak values of different speed intervals.

**Table 1**. Nominal acceleration peaks of the three speed intervals.

| | Speed interval / rad $s^{-1}$ | | |
|---|---|---|---|
| | 157 - 209 | 209 - 261 | 157 - 209 |
| **Acceleration peak/ rad $s^{-2}$** | 166.39 | 202.62 | 241.36 |

#### *3.2.2. Peak variation among different data-processing methods*

In the same speed range, three consecutive measurements of acceleration ramps were performed. For the original mass of data, different signal-processing methods were implemented. Figure 6 shows the behavior that determines the peak in the first acceleration ramp of each method.

Table 2 shows the average peak in each method and the dispersion (standard deviation) among the three measurements. In this case, the results are shown for the intervals of 157-209 rad $s^{-1}$ and 209-261 rad $s^{-1}$.

In the first interval, the standard deviation of the processed data is better than that calculated from the original measurements. The second speed range curve has a lower derivation because the original measurements have better stability.

It is worth noting that in the first interval, the application of the SG smoothing method in both speed and acceleration signals produced identical results for the mean and standard deviation.

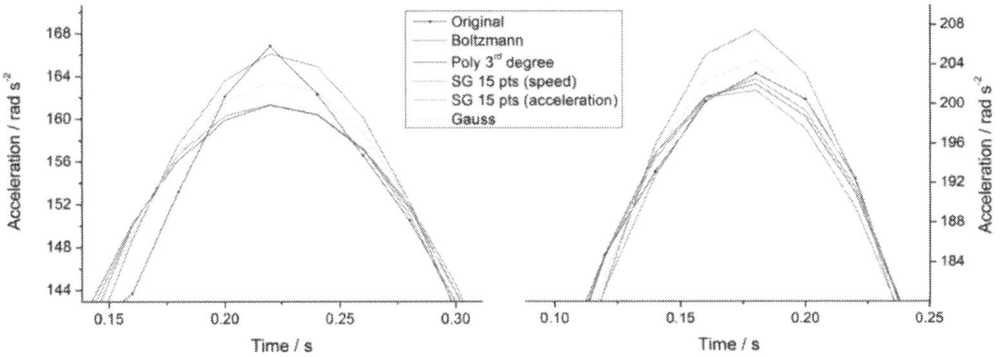


**Figure 6.** Acceleration peaks for different data-processing methods in the speed intervals of (left) 157-209 rad $s^{-1}$ and (right) 209-261 rad $s^{-1}$.

**Table 2**. Mean and standard deviation of the acceleration peaks for the processing methods in different speed intervals.

| Data processing method | Speed interval 157 - 209 rad $s^{-1}$ Average peaks / rad $s^{-2}$ | Speed interval 157 - 209 rad $s^{-1}$ Standard deviation / % | Speed interval 209 - 261 rad $s^{-1}$ Average peaks / rad $s^{-2}$ | Speed interval 209 - 261 rad $s^{-1}$ Standard deviation / % |
|---|---|---|---|---|
| **Original** | 166.39 | 0.29 | 202.62 | 0.25 |
| **Boltzmann** | 166.14 | 0.16 | 207.17 | 0.16 |
| **Poly 3rd degree** | 161.42 | 0.080 | 202.00 | 0.13 |
| **SG 15 pts (speed)** | 161.29 | 0.052 | 202.34 | 0.078 |
| **SG 15 pts (acceleration)** | 161.29 | 0.052 | 201.56 | 0.33 |
| **Gauss** | 163.56 | 0.14 | 204.11 | 0.14 |

### *3.3. Evaluation of acceleration rates*

The acceleration rates correspond to the variation of acceleration in time. This parameter is particularly important for applications that require torque as the control variable. Higher acceleration rates are more difficult to control, and better modeling can better model the control to predict these phenomena. The highest acceleration occurs on the ascending and descending sides of the acceleration ramps. The acceleration rates were calculated from the acceleration curves, and their maximum and minimum correspond to the peaks and valleys. Figure 7 shows the appearance of these curves for each data-processing method. The positive rate indicates an increasing acceleration (pre-peak), and the negative rate indicates a declining acceleration (post-peak).

To quantify these differences, table 3 shows the calculation of the standard deviation of peak rate with and without the methods that were used in two speed ranges.

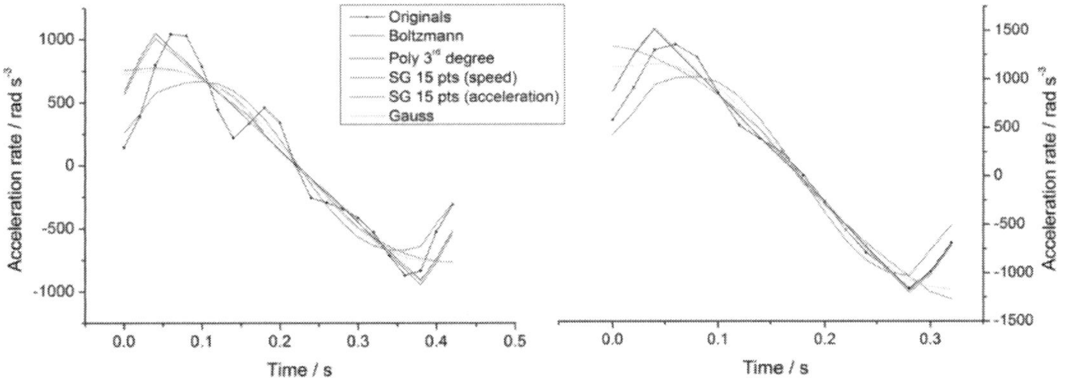


**Figure 7.** Acceleration rates for different data-processing methods in the speed intervals of (left) 157-209 rad $s^{-1}$ and (right) 209-261 rad $s^{-1}$.

**Table 3**. Mean and standard deviation of the peaks of acceleration rates for the processing methods in different speed intervals.

| Data-processing method | Speed interval 157 - 209 rad $s^{-1}$: Average peaks / rad $s^{-3}$ | Speed interval 157 - 209 rad $s^{-1}$: Standard deviation / % | Speed interval 209 - 261 rad $s^{-1}$: Average peaks / rad $s^{-3}$ | Speed interval 209 - 261 rad $s^{-1}$: Standard deviation / % |
|---|---|---|---|---|
| **Original** | 1123 | 7.4 | 1386 | 2.0 |
| | -888 | 2.9 | -1196 | 5.1 |
| **Boltzmann** | 672 | 0.88 | 1021 | 0.36 |
| | -671 | 0.65 | -1020 | 0.74 |
| **Poly 3rd degree** | 1042 | 3.2 | 1493 | 3.1 |
| | -934 | 3.5 | -1207 | 6.5 |
| **SG 15 pts (speed)** | 1018 | 1.6 | 1484 | 2.7 |
| | -958 | 1.5 | -1226 | 5.3 |
| **SG 15 pts (acceleration)** | 821 | 9.9 | 1375 | 2.2 |
| | -773 | 9.8 | -1306 | 3.4 |
| **Gauss** | 771 | 1.3 | 1176 | 1.3 |
| | -772 | 1.2 | -1177 | 1.0 |

### *3.4. Evaluation of different acquisition rates*

A study was also conducted to evaluate the results of the same phenomenon measured at different acquisition rates. Table 4 shows the mean acceleration and deceleration rates and the standard deviation among the three ramps in each acquisition rate. Because of the operational limitations of the experiment, the experiment was only performed for the speed range of 261-313 rad $s^{-1}$. Figure 8 also illustrates different responses for acceleration and deceleration rates with the acquisition rates of 30 Hz and 300 Hz.

**Table 4**. Mean and standard deviation of the peaks of acceleration and acceleration rates for different acquisition rates in the speed interval of 261-313 rad $s^{-1}$.

| **Acquisition rates** | **Acceleration peaks** | | **Peaks of positive acceleration rates** | | **Peaks of negative acceleration rates** | |
|---|---|---|---|---|---|---|
| | Mean/ rad $s^{-2}$ | Standard deviation/ % | Mean/ rad $s^{-3}$ | Standard deviation/ % | Mean/ rad $s^{-3}$ | Standard deviation/ % |
| **300 Hz** | 241.51 | 0.55 | 2244 | 0.93 | -2567 | 1.5 |
| **100 Hz** | 244.68 | 1.26 | 2164 | 8.3 | -2526 | 2.2 |
| **50 Hz** | 241.16 | 0.82 | 2115 | 5.2 | -2409 | 3.9 |
| **30 Hz** | 239.95 | 0.35 | 1876 | 0.50 | -2081 | 0.10 |

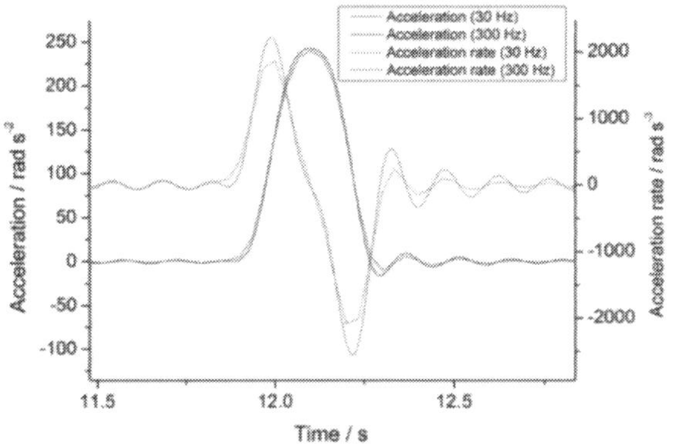


**Figure 8.** Acceleration curve and acceleration rate curve for different acquisition rates.

### *3.5. Data evaluation of torque*

This study measured the torque to legitimize the principle of estimation of the inertial torque using the variables of equation (1). This inertial torque curve was estimated by multiplying the moment of inertia of the measurable portion of the transducer ($\Theta_m$) by the estimated acceleration in the speed ranges. According to the data sheet of the transducer, $\Theta_m$ is approximately 0.00792 kg $m^2$ [7].

Figure 9 shows the torque curves that were obtained from the transducer-measured data and the estimated torque using $\Theta_m$ and the speed data from the adjusted values, which were obtained using the SG method of 17 points.

Although there is a noticeable shift in both speed ranges, the curves have similar behaviors and no intersection. The changes during the peak acceleration because of the change in speed range also occur to the same extent for the variations of the peaks of the torque that was measured using the transducer.

Table 5 shows the results of peak torque and peak torque rate for each torque curve in different speed ranges. As with the acceleration rates, the torque rates were calculated based on the torque curves.

### *3.6. Measurement uncertainty*

It is always important to evaluate the uncertainty of measurement for the proposed system or principle realization. Within the dynamic approach for metrology, in general, the traditional method of calculating the measurement uncertainty based on the ISO GUM [8] becomes incomplete when the magnitudes vary over time, the data are auto-correlated, and the assessment regimes are not stationary [9].

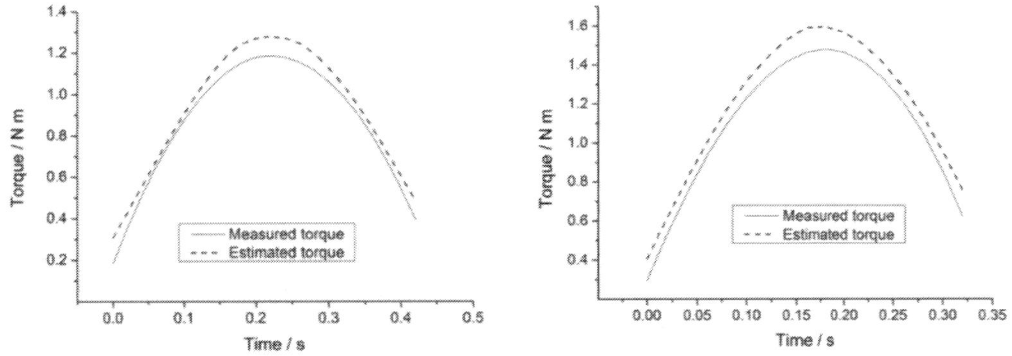


**Figure 9.** Measured torque and estimated torque curves in the speed intervals of (left) 157-209 rad $s^{-1}$ and (right) 209-261 rad $s^{-1}$.

**Table 5.** Mean and standard deviation of the acceleration peaks and acceleration rates for different acquisition rates in the speed interval of 261-313 rad $s^{-1}$.

| | Speed interval 157 - 209 rad $s^{-1}$ | | Speed interval 209 - 261 rad $s^{-1}$ | |
|---|---|---|---|---|
| | Torque peaks/ N m | Torque rates/ N m $s^{-1}$ | Torque peaks/ N m | Torque rates/ N m $s^{-1}$ |
| **Torque Measured** | 1.18 | 8.43<br>-7.53 | 1.48 | 11.83<br>-11.72 |
| **Torque Estimated** | 1.28 | 6.14<br>-6.06 | 1.59 | 10.62<br>-10.06 |

Then, a model is required to balance and consider these issues, which are particularities of dynamic processes, whether they are in the methodology of measurement, adjustment of the involved equipment or data-processing methods [10].

In the case of estimating the inertial torque, the evaluation of the data-processing method is one of the parameters that contribute to the measurement uncertainty. In the approach of modeling the inertial torque curves that are obtained when the shafts are accelerated by the motor drives, the contributions of measurement uncertainty follow the diagram in figure 10.

The dashed lines denote the uncertainty parameters in the validation of the data-processing method, where the repeatability of the parameters under various settings and the errors of indication or interpolation between the original acceleration signal and torque values were discussed in previous sections of this study.

Other contributions to the method validation such as the torque transducer uncertainty, encoder uncertainty and timer (clock) uncertainty of the acquisition system should be considered.

It is important to note that the method must be validated in a controlled environment with well known drives, where it is possible to simulate, evaluate and isolate the parameters of the curves to be studied or measure at the time of application and practical use of these devices, as shown in the previous sections.

However, in the usual applications of the inertial torque estimation, i.e., in the principal use of a shaft in an industrial process, other parameters should be considered and added to the uncertainty of the estimated torque, such as the encoder uncertainty, digital amplifier uncertainty, estimated moment of inertia for the shaft and whenever possible, spot assessment of the repeatability and stability of the acceleration curves.

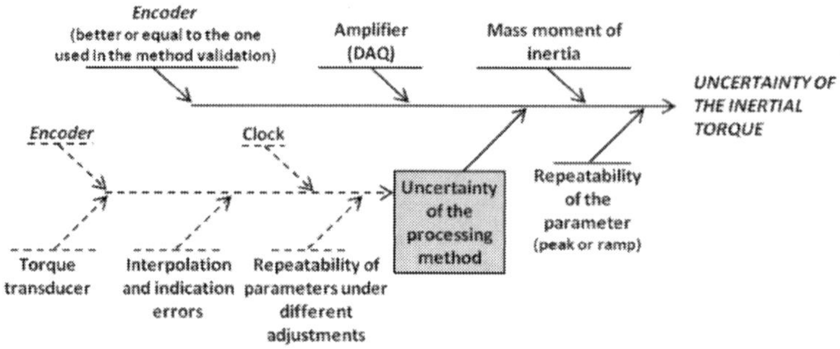


**Figure 10.** Diagram of the sources of measurement uncertainties in estimating the inertial torque.

## 4. Conclusion

Some issues were introduced regarding the necessity and applicability of the estimated inertial torque in driven rotating shafts. The physical principle adopted for this estimate using the acceleration and moment of inertia was presented and evaluated. In the acceleration curves that were obtained from the measurement practices, the regions of interest for the torque evaluation were defined, such as the slopes and peaks.

Because there are two dynamic regimes of measurement, the concept of dynamic error was introduced, and the study focused on evaluating some signal-processing methods of the involved acceleration and speed data, which should smooth these components.

These methods were compared using the repeatability and stability curves, where values were obtained to characterize the curve parameters, these measurement dispersions and the degree of consistency among the methods.

An introductory and qualitative approach was introduced to obtain the main contributions of measurement uncertainty of both the validation methodology of data processing and the inertial-torque estimation in its practical application.

Because this study aims to build a methodology to evaluate the equipment and concepts, it is concluded that although the results demonstrate the feasibility of applying the principles, the proposed foundations must be strengthened, and measurement practices were magnified by adding new equipment and quantification of other parameters involved, particularly those that were detected as uncertainty contributions, which were not addressed here. The torque transducer has a static calibration for its reference measured data, and this lack of dynamic traceability is also an object of parallel study by the authors.

## Acknowledgments

The authors would like to thank the staff of the Motors Testing Laboratory of the Centro de Pesquisas em Energia Elétrica (CEPEL/Adrianopolis-RJ) for their support in performing the measurement practices, CNPq for the scholarships for scientific initiation and the Research Program Bragecrim (Capes/DFG) for the financial support and technical-cooperation viability.

## References

[1] Ovaska S J , Valiviita S 1998 Angular acceleration measurement: a review IEEE Transactions on Instrumentation and Measurement.
[2] Gianfelici F 2005 A novel technique for indirect angular acceleration measurement CIMSA - IEEE International Conference on Computational Intelligence for Measurement Systems and Applications.

Journal of Physics: Conference Series **648** (2015) 012019 doi:10.1088/1742-6596/648/1/012019

[3] Capes/Bragecrim 2012 Research Project: DINTOR - Fundamentals for the Realization and Measurement of Dynamic Torque.
[4] Website: http://www.ptb.de/cms/en/fachabteilungen/abt8/fb-84/ag-842/dynamischemessungen-842.html (jun/2014).
[5] WEG "Motores Elétricos - Guia de Especificação".
[6] Tutorial of the software OriginPro 8.5.0 SR1.
[7] HBM Data-sheet of the transducer T10F.
[8] BIPM, IEC, IFCC, ILAC, ISO, IUPAC, IU-PAP e OIML 2008 Evaluation of measurement data – Guide to the expression of uncertainty in measurement - GUM - JCGM 100.
[9] Martins M A F, Kalid R 2010 Metodologia para avaliação da incerteza de medição em regime dinâmico de sistemas contínuos, COBEQ.
[10] T Bruns, M Kobusch 2005 Data acquisition and processing for PTB's impact force standard machine IMEKO.

# A theoretical and experimental study of the deformation of the piston-cylinder unit of a metrological pressure balance

**L Soares Júnior, F I da Silva Júnior and P Lamary**
Federal University of Ceará, Department of Mechanical Engineering, Fortaleza, CE, Brazil
E-mail: lsjota@gmail.com

**Abstract.** Pressure balance is a measuring system widely used in metrology as a reference standard. The common procedure is to estimate the pressure from an analytical formula involving a set of phenomenological parameters. These parameters are determined experimentally by calibration tests of the balance. In particular, the piston-cylinder deformation is taking into account by the use of a global distortion coefficient and by the use of an effective area of the piston. In order to analyze the possible influence of local deformation on the pressure estimation, a Finite Element (FE) model has been developed using OpenCavok FE code. For a given position of the piston in the cylinder chamber, for a given initial gap between of the cylinder and the chamber, a coupled fluid-structure calculation is carried out. From the local fields, mean values are calculated from which the effective area of the piston-cylinder unit. The analysis is conducted for several engagement lengths of the piston in the chamber and for several loads. The first results obtained show that the local deformation play a role and that FE models could help for precise calculation of the measured pressure.

## 1. Introduction

This paper focuses on pressure balance system used in metrology laboratories as a reference standard to measure pressure. This measurement system is also referred as dead-weight testers. Its operating principle lies in the balance, or compensation, between i) an upward force, applied on the base of the piston and defined as the multiplication of the pressure exerted by the fluid by the known cross sectional area of the piston-cylinder assembly, and ii) a downward force generated by a set of standard masses which act on the top of the balance piston under the influence of local acceleration of gravity [1, 2], as shown in figure 1.

The vertical piston of figure 1 can move freely in the cylinder chamber owing to a micro-metric gap in-between the two parts. This clearance is necessary, so that, while the piston moves, the frictional forces between the two elements are minimized avoiding a possible jamming owing to the forces generated by the action of static friction [3]. The geometry and size of the set are controlled with care and realized with high quality materials, in such a way that the effective area of the piston is determined with high accuracy. The action of the fluid on the base of the piston and on the lateral surface along the cylinder causes mechanical deformation of the assembly. An excessive leakage could alter the effective area depending of the applied pressure value.

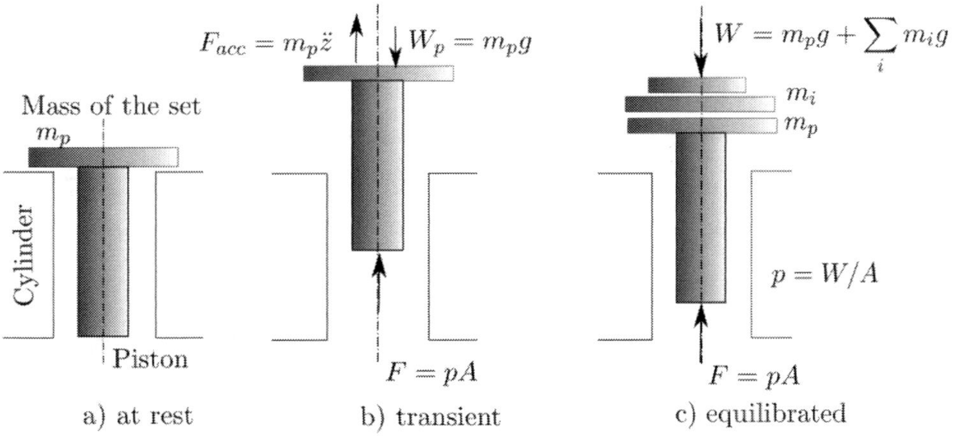


**Figure 1.** Pressure balance measuring principle: a) the balance is at rest, b) transient phase when the balance is put under pressure, c) disks of variable masses are put on the top of the balance as to find an equilibrated position.

The pressure measured by a balance is obtained from the analysis of the different components of the forces applied to the system. For a balance which operates with liquid, the measured pressure is given by:

$$p = \frac{\left[m_p(1 - \frac{\rho_a}{\rho_{mp}}) + m_r(1 - \frac{\rho_a}{\rho_{mr}})\right] g_l + \sigma C}{A_{0,T}\left[1 + (\alpha_c + \alpha_p)(\theta - 20)\right](1 + \lambda p_n)} \pm \rho_{fluide} g_l \Delta h \tag{1}$$

where,
$p$ is the measured pressure, given in $Pa$,
$m_p$ is the mass of the piston, in $kg$,
$\rho_a$ is the specific volumetric mass of the air, in $kg/m^3$,
$\rho_{mp}$ is the volumetric mass of the piston material, in $kg/m^3$ ,
$m_r$ is the sum of the remaining masses operating on the system, in $kg$ ,
$\rho_{mr}$ is the specific volumetric mass of the remaining masses, in $kg/m^3$,
$g_l$ is the acceleration due to the local gravity, in $m/s^2$,
$\sigma$ is the surface tension of the fluid, in $N/m$ ,
$C$ is the length of the circumference of the piston, in $m$,
$A_{0,T}$ is the area of the piston cylinder assembly, in $m^2$,
$T$ is the reference temperature of the piston, in $^oC$ ,
$\alpha_c + \alpha_p$ is the coefficient of the linear thermal expansion of the piston cylinder assembly, in $^oC^{-1}$,
$\theta$ is the temperature of the piston at the time of the measurement, in $^oC$,
$\lambda$ is the distortion coefficient of the piston-cylinder assembly, generally in $(10^{-6})MPa^{-1}$,
$p_n$ is the rated pressure of the measurement given in $MPa$,
$\rho_{fluide}$ is the specific volumetric mass of the fluid used, in $kg/m^{-3}$,
$\Delta h$ is the gap between the piston and the cylinder at the location where the pressure is measured, and is given in $m$.

A way to use equation (1) can be illustrated as follows. Be an equipment, such as a manometer, to be calibrated using the pressure balance. The manometer is connected to the balance and the procedure of figure 1 is followed, putting masses on the top of the balance as to reach the equilibrium of the piston. Next, the formula is used where $p_n$ is the pressure indicated

Journal of Physics: Conference Series **648** (2015) 012020 doi:10.1088/1742-6596/648/1/012020

by the manometer and $p$ is the true pressure.

In equation (1), it can be seen that the variation of the effective area of the piston-cylinder assembly (the denominator of the first term) is assumed linear in regards to the applied pressure. Assuming that the temperatures at the time of the calibration of the pressure balance and at the time of the measurement of an equipment are $20^oC$ ($T = \theta = 20^oC$), the effective area, $A_{effective}$, may be written in the form of equation( 2) as,

$$A_{effective} = A_{0,20}(1 + \lambda p_n) \quad (2)$$

The objective of this work is to study the piston-cylinder assembly behavior from the point of view of the elastic deformations. The effective area given by the phenomenological equation (2) will be compared to numerical simulation based on Finite Element Analysis (FEA). The model and the developments carried out in OpenCavok FE code are also addressed.

## 2. Methodology

The study was conducted considering the VEB Gertewerk oil-operated pressure balance of figure 2. It consists of piston-cylinder unit of figure 3, that can be sighted in the middle of the figure 2 with its initial mass disk on the top. At the left-side and right-side, the balance is equipped with of two vertical tubes where the equipment to be calibrated can be mounted. The balance operates in the nominal range of $[5; 50]MPa$. The hydraulic fluid is a Tellus VG 32 oil.

**Figure 2.** The VEB Gertewerk pressure balance studied.

**Figure 3.** The piston-cylinder assembly.

### 2.1. Calibration tests

The most common method of pressure balance calibration involves a comparison with a second reference pressure balance generating a known pressure. The process operates by cross-floating pressures in a closed system loop and makes the calculation of the effective area (of the balance under test) at a zero-pressure point and at standard temperature of $20^oC$. The pressure distortion coefficient ($\lambda$) of the piston-cylinder assembly is also evaluated [2, 4]. Calibration tests follow a strict procedure that gives rise to calibration test reports. Five calibration tests were carried out over the past 25 years. The table 1 is extracted from these tests and recalls the values obtained for $A_{0,20}$ and $\lambda$, the two parameters involved in the formula of equation (2) under our present interest.

**Table 1.** Measured values of $A_{0,20}$, the area of the piston-cylinder assembly and $\lambda$, the distortion coefficient of the piston-cylinder assembly, from 5 calibration tests.

| | Test 1 | Test 2 | Test 3 | Test 4 | Test 5 | Units |
|---|---|---|---|---|---|---|
| $A_{0,20}$ | 9.971 | 9.994 | 9.989 | 9.984 | 9.984 | $(10^{-6})m^2$ |
| $\lambda$ | — | — | −2.107 | — | −2.107 | $(10^{-6})MPa^{-1}$ |

### *2.2. Finite element model*

A finite element model has been developed using OpenCavok FE software [5]. The Reference FE generator of OpenCavok has been extended in order to be able to define axisymmetric FEs. The related theoretical developments are given in Appendix A.

The finite element model includes two types of physical domain, i) the solid domain, comprising the piston and the cylinder, and ii) the fluid domain, comprising the fluid cavity below the piston and the gap between the cylinder and the piston. The figure 4 presents these parts. The pressure balance is axisymmetric (geometry and loads) and therefore it only needs to be modeled in the half-plane given by the axis of symmetry and a positive radial direction. Bi-dimensional structural and hydrostatic fluid elements were developed for this purpose and used. The localization of the boundary conditions can also be sighted in this picture (figure 4). The axial displacements of the cylinder at the top are restrained. The fluid, also modeled using displacement elements, is contained at the top in the axial direction. The piston is put in a given position for the calculation and restrained at the top for its axial displacement. Along the axis of symmetry, the radial displacements were fixed to ease the resolution. At the bottom, the pressure will be input using a special pressure-to-force axisymmetric mono-dimensional FE.

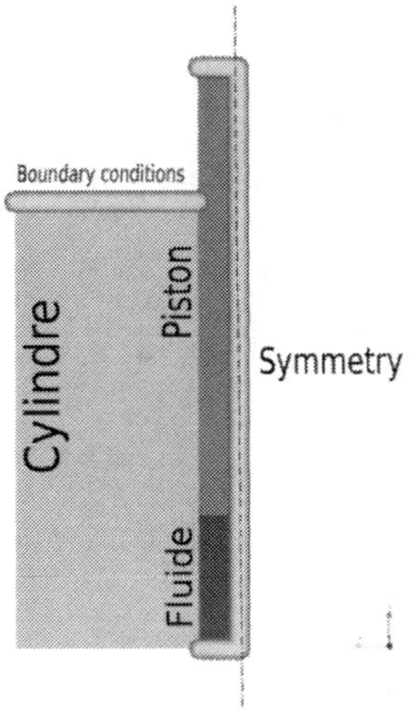


**Figure 4.** OpenCavokV6 axisymmetric finite element model. Main parts and boundary conditions.

**Figure 5.** Zoom on the mesh at the junction of the piston (red) with the cylinder (gray). The fluid cavity (blue) and the fluid contact layer (blue) between the piston and the cylinder can be seen.

Figure 5 is a zoom on the mesh region of the interface between the piston and the cylinder.

Journal of Physics: Conference Series **648** (2015) 012020 doi:10.1088/1742-6596/648/1/012020

The fluid interface is modeled by 3 fluid elements in the radial direction. The size ratio (length/thickness) of these elements is large but their strict rectangular shape should avoid numerical problems. We may note that other approaches are possible. In the research proposed by [6], the fluid is modeled using an analytical formula. The interesting point in this approach is that the fluid model includes more parameters, such as the viscosity and a net flow rate, than FE hydrostatic fluid. The FE displacements of the piston and of the cylinder are reached, in this case, by iteration.

Finally, the model consists in 280,000 finite elements. For a given pressure and a given position, the model is solved in less than 2 minutes. Calculations were carried out considering several positions of the piston and several loads. The analyses took benefit of the full parametrization capability of the FE software.

The data used are the following, the piston has a diameter of $3.5616mm$ and is made of stainless steel with an elastic modulus of $193GPa$ and a Poisson's ratio of 0.31. The cylinder is made of the same material and have a clearance with the piston of $0.003mm$ $(\phi_{cyl} - \phi_{piston})$. For the calculation were considered, a cylinder external diameter of $20mm$, a piston length of $20mm$ and a cylinder length of $20mm$ . The fluid is considered incompressible and a value of $2.54(10^9)Pa$ is used for its bulk modulus.

Concerning the analytical formula of equation (2) that will serve to compare the results, the following data were used: an average effective area of $9.9712(10^{-6})m^2$ and a distortion coefficient of $2.107(10^{-12})Pa^{-1}$.

## 3. Results

The figures 6 and 7 illustrate the results for a given pressure of $p = 35MPa$. Figure 6 is a view of the radial displacement field of the piston-cylinder assembly for a specific engagement length (70%) of the piston in the chamber. Figure 7 is a view of the axial displacements of the piston to verify that the piston moves correctly in that direction.

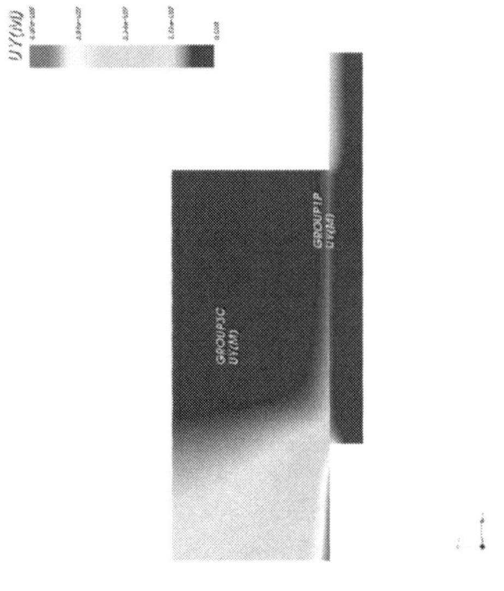

**Figure 6.** Radial displacements, axisymmetric model.

**Figure 7.** Axial displacements, axisymmetric model.

From the nodal displacements of the cylinder and of the piston at their interface, is computed, for each discrete position along the axis, the corresponding effective area. The mean value of all these areas is assumed to be the effective FE area. Figure 8 presents, for a given pressure, the evolution of the effective area while considering several engagement lengths of the piston in the chamber. An engagement length of zero corresponds to the top position of the piston. The results indicate that the effective area is not constant as the engagement length changes. Figure 9 is an analysis for a given engagement length (30%) but with varying input pressure (in the $[5;50]MPa$ range of the pressure balance). It shows that, for this given engagement length, the distortion coefficient (the slope of the curves) is higher when using FE simulation. Both curves, the analytical one, and FE one, are linear.

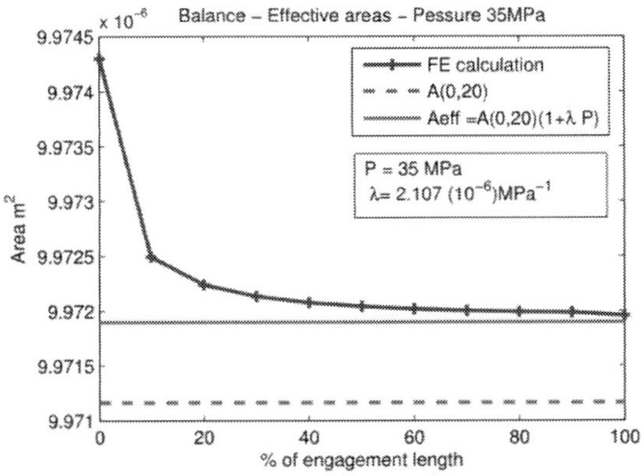


**Figure 8.** Effective areas as a function of the engagement length.

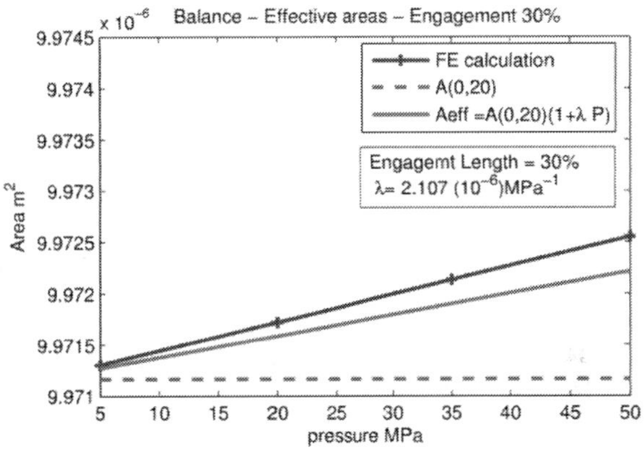


**Figure 9.** Effective areas as a function of the applied pressure.

## 4. Conclusion

Preliminary results from FEA simulations suggest a possible influence of the elastic behavior of the piston-cylinder assembly under various conditions of pressure and piston positions. It seems worthwhile to continue this investigation. From our experience, the structural parts are well-modeled by the equations and by the FE elements used. Coming efforts should concentrate themselves on the fluid models and on the boundary conditions.

## Acknowledgments

The authors would like to acknowledge the CT-UFC, Center of Technology of the Federal University of Ceará, for its financial support and the group the LAMETRO-UFC, the Laboratory of Mechanical Metrology at UFC.

## Appendix A. Axisymmetric FE elements

### *Appendix A.1. General formulation*

The local equilibrium, for any point, $\mathcal{P}$, of the solid domain, $\mathcal{D}$, can be written as the divergence of the stress tensor, $\overline{\overline{\sigma}}(\vec{u})$, $\vec{u}$ being the displacement vector, in the following form,

$$\vec{div}(\overline{\overline{\sigma}}(\vec{u})) = \vec{0} \quad \forall \mathcal{P} \in \mathcal{D} \tag{A.1}$$

The frontier $\partial\mathcal{D}$ of $\mathcal{D}$ can be shared into two parts, $\partial\mathcal{D} = \partial\mathcal{D}_u \bigcup \partial\mathcal{D}_T$, with $\partial\mathcal{D}_u$ a frontier where the $\vec{u}$ is known, and, $\partial\mathcal{D}_T$ a frontier where the normal stress vector is known. We may write these boundary conditions as,

$$\left\{\begin{array}{lcl}\vec{u} & = & \vec{u_d} \quad \forall\mathcal{P}\in\partial\mathcal{D}_u\\ \bar{\bar{\sigma}}\vec{n} & = & \vec{T_d} \quad \forall\mathcal{P}\in\partial\mathcal{D}_T\end{array}\right. \tag{A.2}$$

Following the finite element method, a test function, $\vec{P}$, is introduced and the equation (A.1) is rewritten into an integral form, $W$, as,

$$W=\int_{\mathcal{D}}\vec{P}.\vec{div}(\bar{\bar{\sigma}})dV=0 \quad \forall\vec{P} \quad \text{kinematically admissible} \tag{A.3}$$

Using the identity formula of equation (A.4), equation (A.3) is rewritten as equation (A.5) which allows the use of the Ostrogadsky theorem formula given in equation (A.6) to establish a weak form of $W$ as stated by equation (A.7).

$$\bar{\bar{\sigma}}:\overline{\overline{grad_s}}(\vec{P})=div(\bar{\bar{\sigma}}\vec{P})-\vec{P}.\vec{div}(\bar{\bar{\sigma}}) \tag{A.4}$$

$$W=\int_{\mathcal{D}}(\bar{\bar{\sigma}}:\overline{\overline{grad_s}}(\vec{P})-div(\bar{\bar{\sigma}}\vec{P}))dV=0 \tag{A.5}$$

$$\int_{\mathcal{D}}div(\bar{\bar{\sigma}}\vec{P})dV=\int_{\partial\mathcal{D}}\bar{\bar{\sigma}}\vec{P}.\vec{n}dS=\int_{\partial\mathcal{D}}\vec{P}.\bar{\bar{\sigma}}\vec{n}dS \tag{A.6}$$

$$W_{weak}=\int_{\mathcal{D}}\bar{\bar{\sigma}}:\overline{\overline{grad_s}}(\vec{P})dV-\int_{\partial\mathcal{D}}\vec{P}.\bar{\bar{\sigma}}\vec{n}dS=0 \tag{A.7}$$

Choosing, among all the kinematically admissible test functions $\vec{P}$, the functions which take a null value on $\partial\mathcal{D}_u$ and applying the known values $\vec{T_d}$ on $\partial\mathcal{D}_T$, the final form of the weak variational integral is,

$$W_{weak}=\int_{\mathcal{D}}\bar{\bar{\sigma}}:\overline{\overline{grad_s}}(\vec{P})dV-\int_{\partial\mathcal{D}_T}\vec{P}.\vec{T_d}dS=0 \tag{A.8}$$

Putting the unknowns, $\vec{u}$, on the left-hand side and the known loads, $\vec{T_d}$, on the right-hand side, the equation (A.8) becomes,

$$\int_{\mathcal{D}}\bar{\bar{\sigma}}(\vec{u}):\overline{\overline{grad_s}}(\vec{P})dV=\int_{\partial\mathcal{D}_T}\vec{P}.\vec{T_d}dS \tag{A.9}$$

Following the Galerkine method, the test functions $\vec{P}$ are chosen as virtual displacements that we may note as $\vec{P}=\delta\vec{u}$, leading therefore to the following equation,

$$\int_{\mathcal{D}}\bar{\bar{\sigma}}(\vec{u}):\delta\bar{\bar{\epsilon}}(\delta\vec{u})dV=\int_{\partial\mathcal{D}_T}\delta\vec{u}.\vec{T_d}dS \quad \forall\delta\vec{u} \quad \text{kinematically admissible} \tag{A.10}$$

where $\delta\bar{\bar{\epsilon}}$ is the strain tensor due to the virtual displacements $\delta\vec{u}$.

*Appendix A.2. Axisymmetric formulation*

For axisymmetric solids, in regards to the geometry and to the applied loads, it is assumed that there is no displacement in the circumferential direction. The integrals of equation (A.10) are transformed into a cylindric coordinates system $(\rho,\theta,z)$ from which the elementary volume $dV$ expressed as $dV=\rho d\rho d\theta dz$. Equation (A.10) becomes,

$$2\pi\int_{\mathcal{P}^+}\rho(\bar{\bar{\sigma}}(\vec{u}):\delta\bar{\bar{\epsilon}}(\delta\vec{u}))d\rho dz=2\pi\int_{\partial\mathcal{P}^+}\rho(\delta\vec{u}.\vec{T_d})dl \quad \forall\delta\vec{u} \quad \text{kinematically admissible} \tag{A.11}$$

Journal of Physics: Conference Series **648** (2015) 012020 doi:10.1088/1742-6596/648/1/012020

where,
$\mathcal{P}^+$ is the 2D domain of the solid when cut by the plane $(\rho, z)$ for $\rho$ positive,
$\partial\mathcal{P}^+$ is the oriented contour of this domain from which $dl$ is an infinitesimal element,
All the entities, the tensors and the vectors, have now to be expressed in the cylindrical coordinates system. Which is done using classical formulae.

The domain $\mathcal{P}^+$ is discretized using bi-dimensional finite elements, the domain $\partial\mathcal{P}^+$, if needed, is discretized using mono-dimensional finite elements. Linear approximations were used. The calculus of the integral can now be done as the sum of all the elementary integrals. These are evaluated numerically using the approximation functions and evaluating also $\rho$ at each gauss point. Equation (A.11) takes the following discrete form,

$$2\pi\delta\vec{U}.([K]\vec{U}) = 2\pi\delta\vec{U}.\vec{F} \quad \forall\delta\vec{U} \quad \text{kinematically admissible} \tag{A.12}$$

where,
$\vec{U}$ accounts for the discretized nodal displacements,
$\delta\vec{U}$ accounts for the discretized virtual nodal displacements,
$[K]$ accounts for the global stiffness matrix,
$\vec{F}$ accounts for the applied nodal loads.

As the equation is valid, $\forall\delta\vec{U}$, the algebraic system of equation (A.13) is equivalent to equation (A.12). This system will be solved by the Finite Element solver, introducing, at that time, the kinematic boundaries conditions.

$$[K]\vec{U} = \vec{F} \tag{A.13}$$

*Appendix A.3. Fluid-structure coupling*

For the present work, we considered that the fluid is contained and that there is no net flow rate. The fluid can be modeled using the same approach as for the structural elements, with displacements as main unknowns. The physical components of stress-strain tensors are adapted using the bulk modulus of the fluid. The viscosity could be introduced, to compute a damping matrix, in the case of dynamic analyses.

Introducing the subscript $s$ for the solid and $f$ for the fluid, the coupled system of equations can be written as,

$$\left\{ \begin{array}{lcl} [K_s]\vec{U_s} & = & \vec{F_s} \\ [K_f]\vec{U_f} & = & \vec{F_f} \end{array} \right. \tag{A.14}$$

As the structural parts and the fluid part are modeled using the same types of unknowns (displacements), the coupling is automatic when structural and fluid elements are in contact at a given node. The stiffness terms coming both from the structure and the fluid are added. In the second member of equation (A.14), the structural loads coming from the action of the fluid on the structure are automatically equilibrated by the fluid loads coming from the action of the structure on the fluid and the sum of these components vanish. Finally, a system of the form of equation (A.15) has to be solved.

$$[K]\vec{U} = \vec{F} \tag{A.15}$$

## References

[1] Dadson R S, Lewis S L and Peggs G N 1982 *The Pressure Balance: Theory and Practice* (London: H.M.S.O.)
[2] Dadson R S, Lewis S L and Peggs G N 2011 *Calibration of Pressure Balances* (EURAMET cg-3)

[3] Lewis S L and Peggs G N 1992 *The Pressure Balance: A Practical Guide to Its Use* (London: H.M.S.O.)
[4] Kobata T 2011 Multiple cross-float system for calibrating pressure balances *Proc. of the 5th CCM and 4th IMEKO TC16* (Berlin)
[5] Lamary P, de Araujo Bezerra R, da Silva Junior F I, Pontes de Deus E and Benallal A 2014 A multi-physic implementation of the finite element method apllied to research on acoustic poro-elastic materials *Proc. of the CILAMCE* (Fortaleza) DOI: 10.13140/2.1.3301.6644
[6] Yagmura L and Bagli E F 2009 Experimental and dimensional characterization of a prototype piston-cylinder unit and validation using finite element analysis (FEA) *Measurement* B **42** 678-684

# Horizontal scale calibration of theodolites and total station using a gauge index table.

**L H B Vieira[1], W L O Filho[1], W S Barros[1]**

[1]Dimensional Metrology Laboratory (Lamed), Mechanical Metrology Division (Dimec), Directory of Scientific and Industrial Metrology(Dimci), National Institute of Metrology, Quality and Technology (Inmetro), Av. Nossa Senhora das Graças, 50, Xerém, Duque de Caxias, RJ, Brazil, ZIP 25250-020

E-mail: lhvieira@inmetro.gov.br

**Abstract.** This paper shows a methodology to calibrate the horizontal scale of theodolites and total station using a high accuracy index table. The calibration pursued the method of circular scales and precision polygons (also called Rosette Method [1] or multistep). This method consists in the angle comparison of two circular divisions in all relative positions possibilities. Index table errors and theodolite horizontal scale errors were obtained using the method of least squares which is used to process the data from Rosette Method. An experimental setup was used to evaluate this methodology and the details of the mechanical assembly are also described in this paper. Several theodolites and total stations were calibrated using the proposed system and the results infer that the method is suitable to calibrate the different models available in the market. The system showed good stability over time with measurements uncertainties around 1" (one second) depending on instrument features.

## 1. Introduction

Theodolite is an instrument primarily used in vertical and horizontal angle measurements. The measurements are carry out by a rotational telescope that moves on both axis. In order to measure the displacement of the telescope, each axis contains a graduated scale as shown in figure 1.

The total station consists in the same principle of theodolite, but some changes can be found such as the incorporation of an electronic distance meter (EDM) and encoders instead of a circular graduated scale. The modern total stations can be operated by remote control or automatically measurements even with the target in movement [2].

These instruments have application in different areas, for example, road construction and aircraft turbines alignment. In the industrial measurement, high accuracy electronic theodolites have been used in automobile manufacturing, shipbuilding and railways constructions, the nuclear industry, the oil industry and robotics.

Generally, the technical standards present methodologies of qualifying and classification of these instruments. However, the calibration concept demands traceability evidences by higher accuracy metrological standards (national or primary standards). Many other calibration methods of theodolites horizontal scales and total stations can be performed based on circular scale and encoder calibration technique [3]. This paper presents a technique that was developed in the National Institute of Metrology, Quality and Technology (Inmetro) to calibrate the horizontal scales of theodolites and total stations. In particular, the reference [2] presents a vertical scales calibration method using a coordinate measuring machine (CMM).

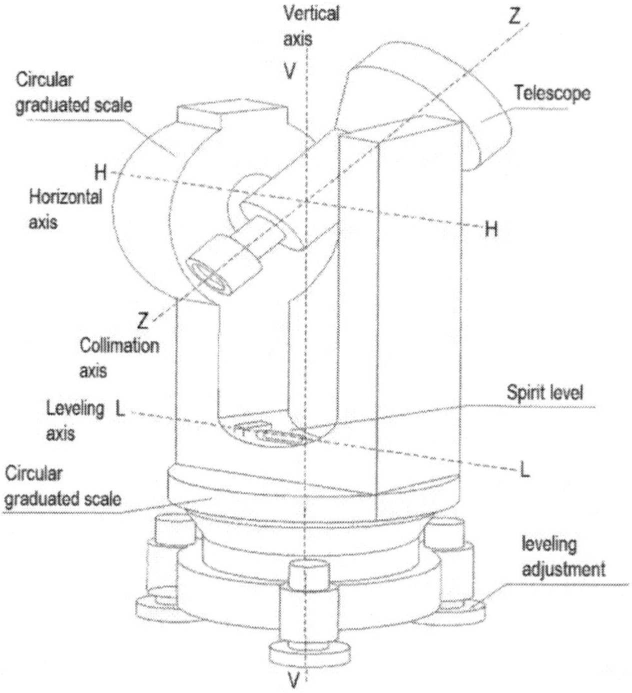


**Figure 1.** Theodolite schematic drawing [4].

## 2. Objective

This paper presents a methodology to calibrate the horizontal scales of theodolites and total with a gauge index table applying a method which is commonly used to perform the angle unit by the comparison of two circular divisions, the Rosette Method.

## 3. Methodology

The calibration method consists in a comparison between two circular divisions. In this case, one circular division is the theodolite horizontal scale and the other one is the index table. The theodolite angles are compared with the index table angle in all possible relative positions and the method is based on circle closing where the sum of partial angles errors is equal to zero. This methodology is self-consistent and can be realized without other standards as reference, allowing the simultaneous calibration of the theodolite scale and the index table.

The experimental setup was assembled on a rigid bench, as shown in figure 2. A rotary table was used to provide the relative position changing in the comparison system. The gauge index table was assembled on the rotary table and the theodolite was mounted onto this setup. Between the theodolite and the gauge index table was installed a levelling platform, which supported fixation and levelling. A collimator was mounted on the rigid bench and aligned relative to the theodolite telescope. This collimator emulates large distances and has a reticle which is used as a reference (target).

**Figure 2.** Measurement system assemble

As the system is composed by several manual mechanisms, it is necessary to be careful with instrument aligning and all other system components. The calibration started with index table rotation to an angle counterclockwise. After that, the theodolite telescope was returned clockwise to the same angle until the collimator reticle coincides with the theodolite reticle. The indicated value in the theodolite horizontal scale represents the index table error subtracted by the theodolite scale error and the alignment error between the two circular divisions ($d = \alpha - \beta - \gamma$). The procedure continued covering all possible relative positions until all data were obtained.

This comparative method generates a system of $n^2$ equations with $2(n - 1) + n = 3n - 2$ unknowns, where $n$ is the number of measurement points of the instrument scale. Therefore, with more equations than unknowns, it is impossible to find exact solutions. The errors of the scale and the index table are estimated using the least squares method [5, 6].

The measured values ($d_{ij}$) are represented by the equation:

$$d_{ij} = \alpha_i - \beta_j - \gamma_k \tag{1}$$

With $i, j$ e $k$, ranging from 0 to ($n$-1).

$$k = \begin{cases} i - j; & \text{if} \quad i - j \geq 0 \\ i - j - n; & \text{if} \quad i - j < 0 \end{cases} \tag{2}$$

Where:

$d_{ij}$ = measured values;
$\alpha_i$ = error of the angles of indexing table;
$\beta_i$ = errors of the angles of the horizontal scale;
$\gamma_k$ = alignment error between the origins of the indexing table angles and horizontal scale.

Journal of Physics: Conference Series **648** (2015) 012021 doi:10.1088/1742-6596/648/1/012021

The solution to $\alpha_i$, $\beta_j$ e $\gamma_k$ are as follows:

$$\hat{\alpha}_i = n^{-1}.(\sum_{j=0}^{n-1} d_{ij} - \sum_{j=0}^{n-1} d_{0j}) \qquad i = 1 \text{ a } (n - 1) \tag{3}$$

$$\hat{\beta}_j = -n^{-1}.(\sum_{i=0}^{n-1} d_{ij} - \sum_{i=0}^{n-1} d_{i0}) \qquad j = 1 \text{ a } (n - 1) \tag{4}$$

$$\hat{\gamma}_k = -n^{-1}\sum_i d_{i,(i-k)} + 2n^{-2}\sum_i\sum_j d_{ij} - n^{-1}\sum_i d_{i0} - n^{-1}\sum_j d_{0j} \qquad k = 0 \text{ to } (n - 1) \tag{5}$$

The estimative of standard deviation (S) of index table and horizontal scale accumulated angles errors of the theodolite is given by:

$$S = S_d.\sqrt{2n^{-1}} \tag{6}$$

$S_d$ is given by:

$$S_d^{\,2} = \frac{\sum_{i=0}\sum_{j=0}(\hat{\alpha}_i - \hat{\beta}_j - \hat{\gamma}_k - d_{ij})^2}{(n-1)(n-2)} \tag{7}$$

Where $(n - 1)(n - 2)$ is the degrees of freedom number, in other words, the number of measurements $(n^2)$ subtracted by the number of unknowns $(3n - 2)$.

The expanded uncertainty of the proposed method is given by the following equation:

$$U = k \cdot \sqrt{u_s^{\,2} + u_r^{\,2} + u_o^{\,2}} \tag{8}$$

Where:

$u_s$= uncertainty due to the standard deviation;
$u_r$= uncertainty due to resolution of the instrument;
$u_o$= uncertainty due to operator error.

**4 Results and discussion**

Two theodolite Wild, T3 model with reading 0.1"; two total stations Berger, CST-202 models and CST-205 both with reading 1" were calibrated. The figure 3 shows the means of the theodolite Wild T3 indication errors, calibrated in 2008.

Systematic errors were found in relation to the eccentricity between the horizontal scale and the vertical axis. The measurement reproducibility was good and the differences between the two calibrations did not exceed 0.6". The calibrations were obtained in a time lapse of one month and showed no incompatibility results, as can be seen in table 1. The measurement uncertainties of these two calibrations were 0.9"and 0.7" for the first and second calibration in 2008, respectively.

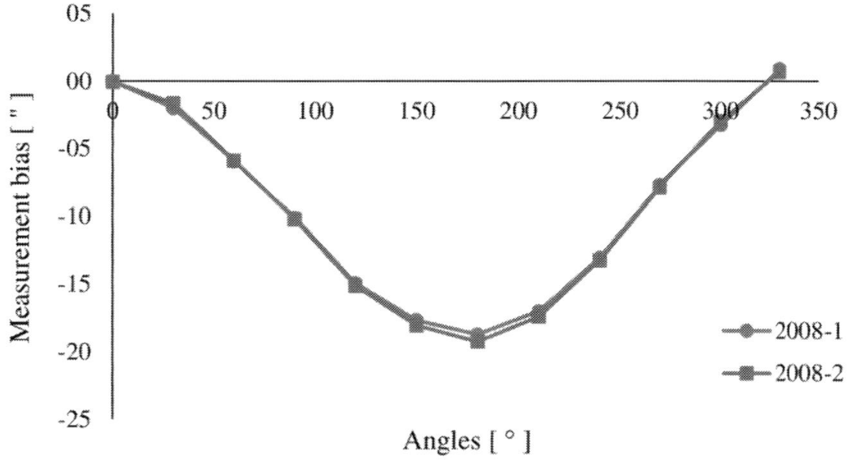


**Figure 3.** Calibration results of theodolite Wild T3 in 2008

**Table 1.** $E_n$ results for Wild T3 calibration.

| Angles ( ° ) | $E_n$ |
|---|---|
| 0 | 0.0 |
| 30 | 0.1 |
| 60 | 0.3 |
| 90 | 0.0 |
| 120 | 0.1 |
| 150 | 0.2 |
| 180 | 0.3 |
| 210 | 0.5 |
| 240 | 0.4 |
| 270 | 0.2 |
| 300 | 0.1 |
| 330 | 0.3 |

The compatibility of calibration results were evaluated by the comparison of measurement errors and expanded uncertainty, pursuing the ISO/IEC 17043 [7]. The Normalized Error ($E_n$) was calculated using the following equation:

$$E_n = \left| \frac{V_{test} - V_{ref}}{\sqrt{U_{test}^2 + U_{ref}^2}} \right| \quad (9)$$

Where:

$V_{test}$= results of object;
$V_{ref}$= results of reference standard;

$U_{test}$= expanded uncertainty of object;
$U_{ref}$= expanded uncertainty of reference standard.

Two values were considered compatible when the $E_n \leq 1$.

Another theodolite Wild T3, was calibrated in 2010 and 2013. The results are shown in figure 4. Only the 60° point revealed an $E_n$ out of the limits as can be seen in table 2. This incompatibility can be justified considering that theodolites are instruments used in the field and submitted to large variations in environmental conditions and collisions. The uncertainties obtained were 0.9" and 0.7" for calibrations done in 2010 and 2013, respectively.

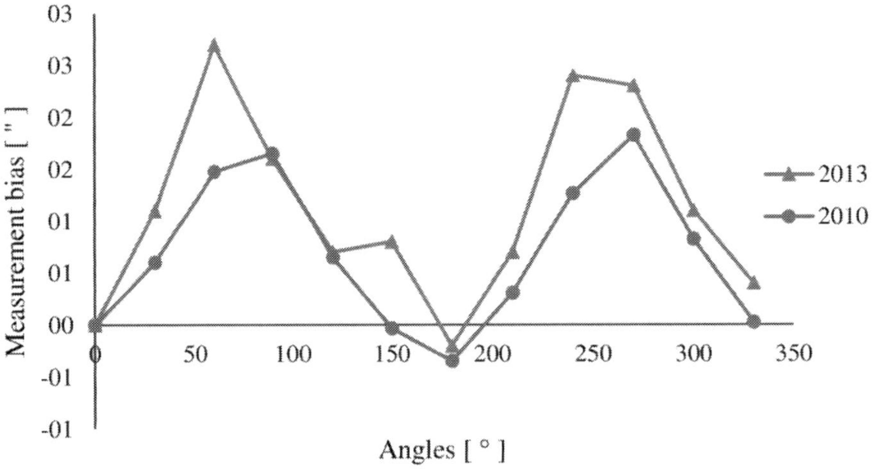


**Figure 4.** Calibration results of second theodolite Wild T3 in 2010 and 2013

**Table 2.** $E_n$ results for second Wild T3 calibration.

| Angles ( ° ) | $E_n$ |
|---|---|
| 0 | 0.0 |
| 30 | 0.4 |
| 60 | 1.1 |
| 90 | 0.0 |
| 120 | 0.0 |
| 150 | 0.7 |
| 180 | 0.1 |
| 210 | 0.4 |
| 240 | 1.0 |
| 270 | 0.4 |
| 300 | 0.2 |
| 330 | 0.3 |

The calibrations of the total station CST Berger 202 was performed in three different conditions and the results are in figure 5. All measurements were performed in the same month in 2009, but the measurements 1 and 2 were carried out by different operators and the measurement 3 was performed using the vertical scale in 270°.

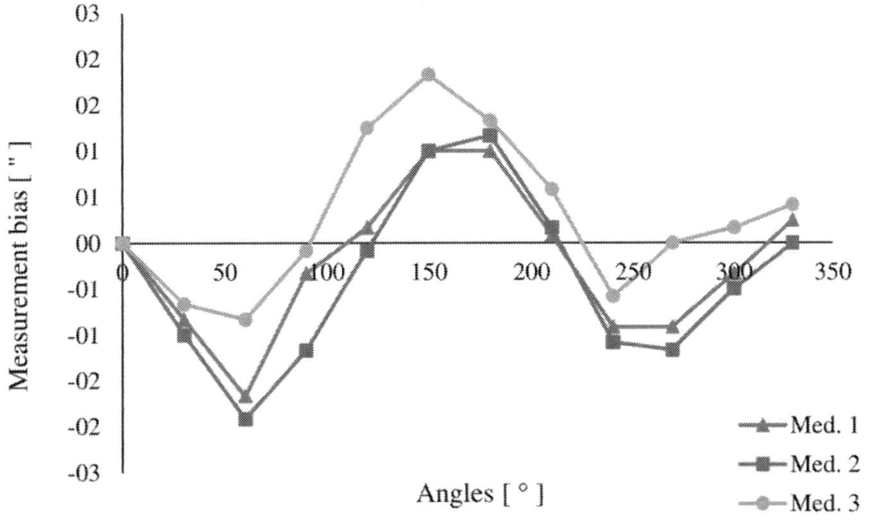


**Figure 5.** Calibrations results of total station CST Berger 202

Although the calibration has been performed in different conditions as described above, the results revealed good reproducibility considering the uncertainties of the method, which were limited by the instrument resolution (1"). There is no incompatibility between the results in these calibrations, as shown in table 3. The measurement uncertainties obtained for the three calibrations were respectively 1.8"; 1.7" and 1.7".

**Table 3.** $E_n$ results for Berger CST 202 calibration.

| Angles (°) | $E_n$ 09-1/09-2 | $E_n$ 9/1/2010 | $E_n$ 9/2/2010 |
|---|---|---|---|
| 0 | 0.0 | 0.0 | 0.0 |
| 30 | 0.2 | 0.0 | 0.2 |
| 60 | 0.2 | 0.3 | 0.4 |
| 90 | 0.2 | 0.8 | 1.0 |
| 120 | 0.2 | 0.3 | 0.5 |
| 150 | 0.0 | 0.2 | 0.2 |
| 180 | 0.2 | 0.5 | 0.3 |
| 210 | 0.6 | 1.3 | 0.5 |
| 240 | 0.5 | 1.6 | 1.0 |
| 270 | 0.4 | 1.2 | 0.7 |
| 300 | 0.2 | 0.8 | 0.5 |
| 330 | 0.2 | 0.4 | 0.2 |

Another Berger total station was calibrated in 2009 and 2010, the model CST station 205. The results of the three calibrations are shown in figure 6. The two calibrations in 2009 were performed by different operators and in 2010 by the operator of first calibration in 2009. A comparison between calibrations was performed and the results are shown in table 4.

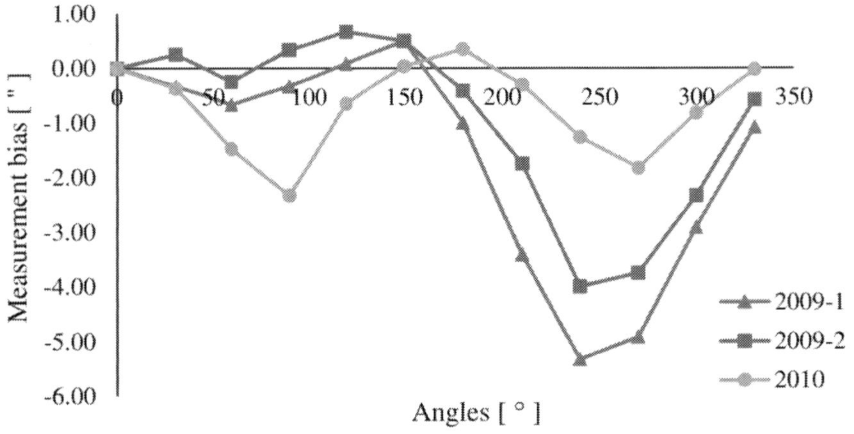


**Figure 6.** Calibrations results of total station CST Berger 205

**Table 4.** $E_n$ results for Berger CST 205 calibration.

| Angles (°) | $E_n$ 1-2 | $E_n$ 1-3 | $E_n$ 2-3 |
|---|---|---|---|
| 0 | 0.0 | 0.0 | 0.0 |
| 30 | 0.1 | 0.1 | 0.1 |
| 60 | 0.1 | 0.3 | 0.5 |
| 90 | 0.3 | 0.1 | 0.5 |
| 120 | 0.1 | 0.4 | 0.6 |
| 150 | 0.0 | 0.3 | 0.3 |
| 180 | 0.1 | 0.1 | 0.1 |
| 210 | 0.0 | 0.2 | 0.2 |
| 240 | 0.1 | 0.1 | 0.2 |
| 270 | 0.1 | 0.4 | 0.5 |
| 300 | 0.1 | 0.2 | 0.3 |
| 330 | 0.1 | 0.1 | 0.2 |

In some points there are variation between the results of the 2010 and 2009. However a variation of approximately 4" still within the tolerance specified by the manufacturer [8]

Analyzing the comparison results, some incompatible values can be found, but may have occurred, in this case, an underestimation of measurement uncertainties.

## 5. Conclusion

The measurement results prove that the proposed method is suitable for performing calibration of theodolites and total stations horizontal scales. The measurement system is easily applicable and is not dependent on the type and size of the different equipment models available on the market.

The measuring uncertainty depends practically of two components: the instrument resolution and the repeatability inherent from data analysis method (least squares method). The operator influence was perceived mainly in the case of analogic theodolites.

## References

[1] OIML 1994 Methods of reproduction of plane angle units *OIML Bulletin* **35** 23-29

[2] Filho W L O, Marques A and Vieira L H B 2011 Método Para Calibração Da Escala Angular Vertical De Teodolitos E Estações Totais Através De Medição Por Coordenadas *2° Cong. Int. de Metrologia Mec*

[3] Bručas D, Giniotis V, and Petroškevičius P 2006 The construction of the test bench for calibration of geodetic instruments Geodezija ir kartografija **32** 66–70

[4] Branco D F 1998 Calibração de Teodolitos *master's thesis from Pontifícia Universidade Católica do Rio de Janeiro*

[5] Vieira L H B 1996 Metrologia dimensional: Realização das unidades de ângulo plano *master's thesis from Universidade Federal Fluminense*

[6] Cook A H 1954 The calibration of circular scales and precision polygons *Br. J Appl. Phy* **5** 367-371

[7] Associação Brasileira de Normas Técnicas (ABNT) 2011 Avaliação de conformidade - Requisitos gerais para ensaios de proficiência. *ABNT NBR ISO/IEC* **17043**

[8] CST/Berger Electronic Total Station User's Guide n.d. Retrieved Jul 06, 2014 from http://www.southern-tool.com/store/media/cst/total-station/56-CST202.pdf

# Experimental evaluation of a facility for jet induced flow analysis

**M H Farias, A M Santos and Y B Zanirath**

Fluid Dynamics Metrology Division, Scientific and Industrial Metrology, National Institute of Metrology, Quality and Technology (Inmetro), Rio de Janeiro, Brazil

E-mail: mhfarias@inmetro.gov.br

**Abstract.** In this work, the performance of a bench, which was designed for investigating the hydrodynamics of jets and plumes in quiescent environment is evaluated. This bench is mainly composed by a water tank with a nozzle placed at the center of its bottom. Through the nozzle, fluid is injected vertically upward into the tank. The tank, with square cross section, has a device attached to its upper edge for containing overflow. For fluid flow measurement and visualization, the optical technique Stereo Particle Image Velocimetry (SPIV) was employed and the influence parameters for the resulting flow pattern inside the tank were analyzed. The investigations showed that, in order to avoid tank wall effects to the axisymmetric jet profile, the minimum and maximum flow rate of injected liquid allowed in the stagnant environment were 0.05 L/min and 0.20 L/min, respectively. All the measurements were accomplished under the maximum water column in the tank.

## 1. Introduction

Jet induced flow is frequently used in industrial or technological processes and it is present in natural events. This kind of flow is highly complex[1]. Although several studies have been done for getting a best understanding on this flow behavior, more still need to be done. Improvements and advances in important areas, such as irrigation, oil and gas industry, environmental (flow in rivers, atmosphere, lake and oceans, as well as pollutant dispersion processes in these environments etc.), among others, can depend strongly on the knowledge of the hydrodynamics phenomena which occurs in jet or plume induced flow. In pursuing reliability of the measurements when detailed investigations of flows induced by jets or plumes are performed, metrological demands arise. Difficulties due to technologies capability limitations in addition to the physical phenomena complexity characteristics of the flow make the work a challenging task. In the literature, issues related to jets and plumes flows [2,3] clearly show that exists a not deeply explored field, this one concerned to improvement of measurement procedures of such complex flows, which claims to metrologist attention. So, the aim of the present work is to identify, considering rigorous metrology requirements, the applicability limits of a facility for studies of mixing process of liquids injected in quiescent liquid. The bench is a tank containing accessory for liquid overflowing storage. As the procedure of evaluating such facility requires investigations of the mixing flow inside the tank, the analysis also includes the velocity field behavior characterization for a turbulent induced flow. The experimental apparatus and accessories were designed and constructed by the Brazilian National Metrology Institute -

Inmetro researchers, whom intend to contribute for knowledgement advances in the complex flow dynamics area.

## 2. Experimental apparatus and setup

### *2.1. The apparatus assembly*

The experimental apparatus was built to analyze the dynamics of liquid mixing process driven by fluid injection in stagnant environment. In this kind of bench, an important metrological requirement which has to be fulfilled is the absence of disturbances generated by sources which are not due to the own fluid injection process. In this direction, during the design, assembling and commissioning of the bench, all scaling and configurations definitions were oriented by the criteria of minimization of undesirable influences on the flow.

The first step of the assembling task was the connection of an acrylic tank to the hydraulic circuitry, as shown in Figure 1. Externally and close to the upper edge of this tank, a device for liquid overflow catching was attached. When liquid jet is inserted into the stagnant liquid, the overflow can occur deliberately if the experiment starts under the maximum liquid column of the tank or if the purpose of the experiment is to observe the free surface in upward movement.

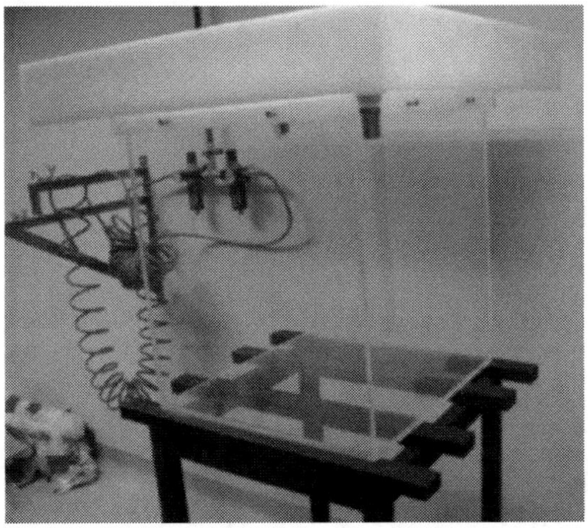

**Figure 1.** Tank with a device for containing overflow

The tank has square section of 0.5 m x 0.5 m and 0.7 m in height, with rounded edges in the top. The rounded edges were introduced with the aim of benefiting the homogeneity of fluid distribution when the liquid overflows. With the tank buttressed, the bench mounting was developed as illustrated in Figure 2. The facility after assembled is shown in Figure 3.

The reservoir was mad e of polypropylene and its dimensions are 0.5 m x 0.5 m x 0.4 m (L x W x H). Through a PVC pipe of inner diameter 1 inch, the suction end of a progressive cavity pump (model HD15, Weatherford$^{TM}$, Geremia) was connected to the reservoir bottom, and in the pump discharge end was installed a valve(bypass) in order to control the liquid return quantity to the reservoir and also into the tank. The work fluid was filtered water, and the flow rate was monitored by an eletromagnetic flowmeter (Conault$^{TM}$). The flowmeter was calibrated in totalized volume quantity, and also in instantaneous volumetric flow rate. The nozzle injector, with capillary length to diameter ratio of 10 (length = 10 mm and inner diameter = 1 mm), was made in 304 stainless steel. All the pipeline have inner diameter of 0.5 inch.

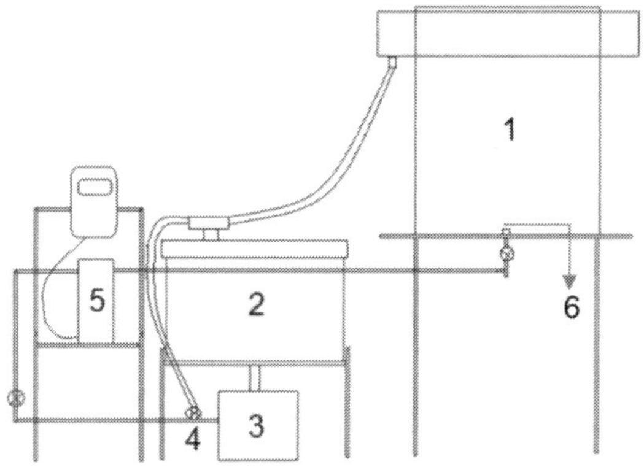


**Figure 2.** Squematic diagram of the experimental apparatus (1: tank, 2: reservoir, 3: pump, 4: bypass, 5: flowmeter, 6: injector)

**Figure 3.** The experimental apparatus

*2.2. Velocity measurements by particle image velocimetry*

In the Particle Image Velocimetry-PIV method, the instantaneous fields of the velocity vector are determined through the displacement of fine particles which move following the fluid motion [5,6]. The PIV technique consists of two pulsed light sources (sheet of light) which illuminate tracer particles in the flow field, and then high speed cameras, CCD or CMOS, acquire sequential images of the scattered light from the particles. By using mathematical tools as cross-correlation or auto-correlation, the displacement field is determined as an average displacement of particles within small regions (called interrogation areas) of the image plane captured during a known time interval between the acquisition of two images [7]. It is possible to determine the velocity vector on 2 or 3 components. In most of literature reference, the PIV technique is considered non-intrusive, but according to metrology concepts, in fact it is intrusive, since the technique demands insertion of particles in the flow.

In this work, the system employed was 3D (StereoPIV) from LaVision™. The system is a double cavity Nd:YAG laser, with power of 120 mJ/pulse at 532 nm, in which two laser sheet pulse synchronized with two high speed cameras aperture. The trigger signal for both Nd:YAG laser is given by the LaVision™ programmable timing unit (PTU), manually or by computer control. The integrated cameras of the system (LaVision™, Imager Pro X 2M, 1648 x 1214 pixel resolution, from 30 frames/s up to 135 frames/s) are based on a charge-coupled device (CCD) sensor with high resolution and high sensitivity. In the experiments, glass filter high pass (cut-off at 540 nm) were installed in each camera, and both cameras captured images in double frame, with inter frame time of 5500 $\mu$s. As tracer, particles in sizes between 20 $\mu$m and 50 $\mu$m of Rhodamine B (PMMA-Rhodamine B aqueous suspension 100 g/400mL) were used.

The system is operated by DaVis™7.2 software, and the particles shift is calculated in the post-processing for determining the velocity vectors and associated turbulence characteristics of the flow field.

### *2.3. Experimental procedure and data processing*

Before using the PIV system for measurement of the resulting flow field characteristics inside the tank, it was necessary to determine the minimum and maximum water flow rate on which the pump of the bench could be operated. Among the acceptance criteria for flow rate operation limits, two were detached: a) the need of stability maintenance for the flow rate along the time and b) stability of the pressure level at the pump exit during the running experiments. For the minimum water flow rate identification, by considering that one critical condition for the pump operation is the minimum value of the motor pump speed on which none fail or stop due to motor temperature increasing could occur, tests towards accomplishing this requirement were performed. In this case, as the worst situation was set up when the bypass was totally closed (see the bypass in Figure 4), the limits were experimentally searched out under this condition.

**Figure 4.** Detail of the bypass

Although the minimum water flow rate had been pursued in accordance to the aforementioned conditions, after to find this value it was important to verify if in such minimum flow rate the PIV system could detect movement of the tracer particles in the induced flow inside the tank. Thus, to make sure the minimum flow rate achieved in the previous step was enough, a comparison between two subsequent images of the flow captured by the PIV cameras was conducted. If the frequency of image acquisition is coherent for getting details of the flow behavior, the reliability of the resulting flow mapped by PIV increases. However, this right frequency is empirically achieved, since it also will depends on the own flow behavior. As the resulting flow mapping quality is affected by the capacity of combining the best image acquisition frequency with planned scale for the flow details detection, in metrological sense it means that the performance of the PIV also plays an important role on the bench applicability limits, if PIV is chosen to be used in

Journal of Physics: Conference Series **648** (2015) 012022 doi:10.1088/1742-6596/648/1/012022

the flow investigations. As in the present article the authors not intend to deep on PIV system settings, the readers can get more details about this subject in the literature.

Another point to consider in the bench performance analysis is the influence of tank wall effects to the jet entrainment rate. Consequently, nevertheless liquid overflows the tank, the flow rate at the nozzle always affects the magnitude of the recirculating zone within the tank. Then, considering all these aspects, after some tests the flow rate levels selected to the bench evaluation were 0.05 L/min and 0.20 L/min. The experiments were performed under the maximum water column (0.70 m), and the rhodamine concentration in the tank and reservoir were equal. Previous experiments carried out in the laboratory were used as base for determining the tracer particles concentration in these experiments.

With the PIV system, a synchronization circuit between the pulsed laser and imaging system allowed the digitalization and storage of double frame images, totalizing 1190 images (7 sets containing 190 images each one) with a time separation of 5500 $\mu$s between frames. After capturing the 1190 images for each flow rate, the post-processing was initiated. So, the camera images were divided into rectangular sections (interrogation regions) in two steps: firstly, into interrogation area equal 64 x 64 pixels and, second, into 32 x 32 pixels. The condition for determining the interrogation area size is that it must be sufficiently small so that the velocity gradients across it can be neglected, but large enough to ensure that there is sufficient seeding particle pair to contribute to the statistical analysis of the fluid movement. In these experiments, the flow visualized region was in a vertical plane placed just above the nozzle exit, 120 mm large and 120 mm in height. The symmetry axis of this plane coincided with the nozzle axis. In this way, it was possible to evaluate the behavior of the liquid jet, and also, to compare such results with the similar case of axisymmetric air bubble plume on which the authors have been studying [7].

Finally, the main variables useful to evaluate the resulting flow field were explored using the free software SciDAVis$^{TM}$ (Scientific Data Analysis and Visualization) supported by Excel spreadsheets from Microsoft$^{TM}$.

## 3. Results and discussion

The obtained results of the axisymmetric jet flow field in the minimum and maximum flow rate are presented in Figures 5 to 8. In Figures 5 and 6, X-axis shows the parallel plane to the tank bottom and Y-axis indicates the vertical position along the tank height.

The vertical velocity distributions, as shown in Figures 5 and 6, along a horizontal section of the tank show a Gaussian profile, and are in accordance to results referred to in the literature [1].

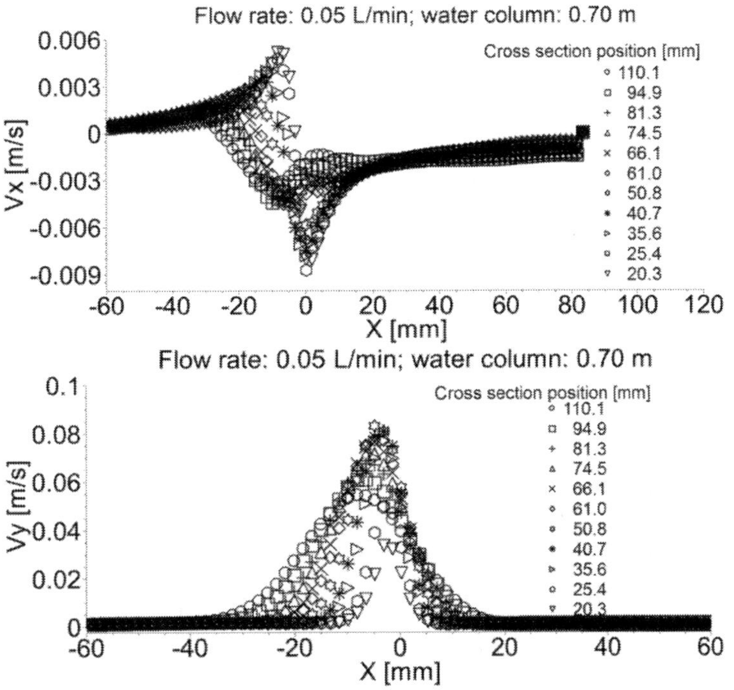


**Figure 5.** Velocity profiles Vx (top) and Vy (bottom) at several horizontal cross section positions for the minimum flow rate

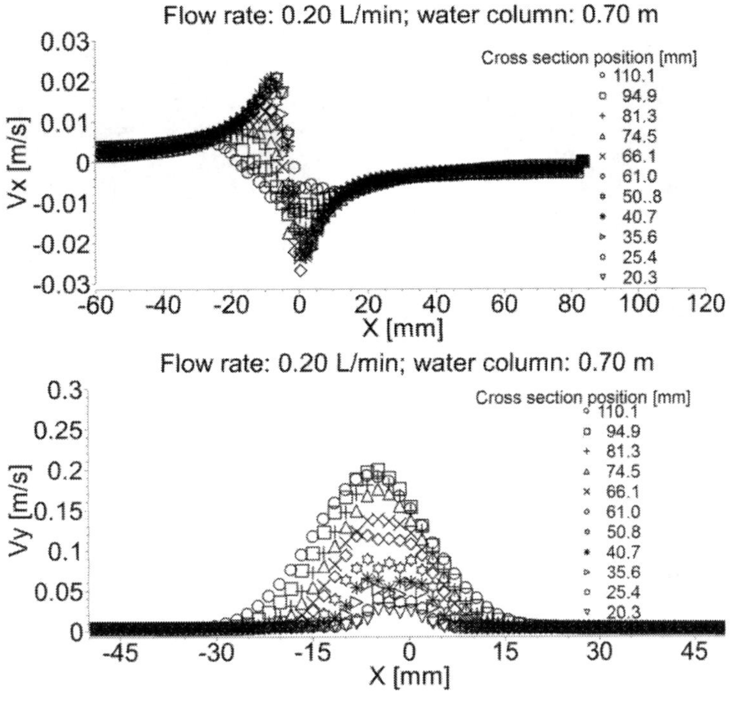


**Figure 6.** Velocity profiles Vx (top) and Vy (bottom) at several horizontal cross section positions for the maximum flow rate

Since the injector is an axisymmetric nozzle, it is expected a coincidence of the symmetry lines of each jet vertical velocity profile on different distances of a horizontal section to the injector. Nevertheless, the peak values of Vy does not coincide with the injector centerline when the flow rate is minimum. Furthermore, on the flow rate equal to 0.05 L/min, near the injection point the maximum velocity Vy increases slightly, but it decreases as the position up to the injector increases. For maximum flow rate (0.2 L/min), the peak velocity (Vy) increases gradually until to reach a plateau near 0.2 m/s, how is shown in Figure 7.

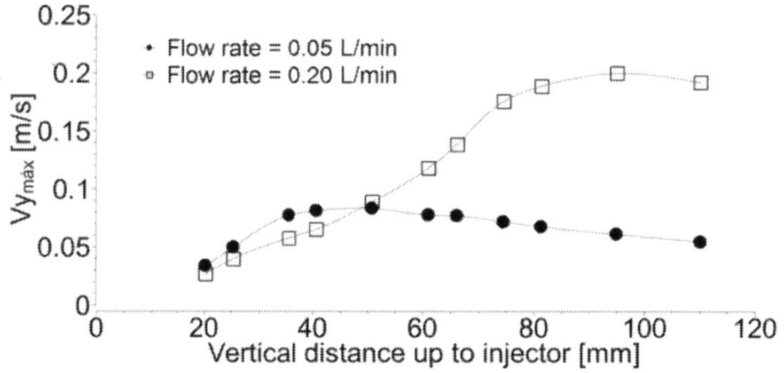


**Figure 7.** Velocity peaks at several horizontal cross section positions

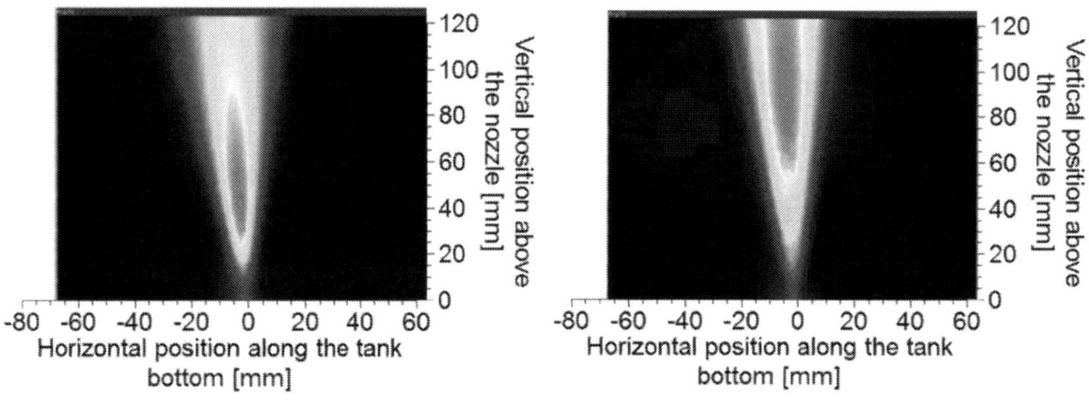


**Figure 8.** Velocities field for the minimum flow rate (left) and the maximum flow rate (right); colors range: near 0 m/s - dark blue, 0.2 m/s - red; graphs generated by DaVis$^{TM}$7.2 and adapted by the authors

Analyzing the Figure 8, it is noted that in case of low flow rate the core of high velocities is situated near the injection point and it occupies a small volume of the axisymmetric jet. Otherwise, for high flow rate this core is larger than for low flow rate and it is set far from the injection point. It occurs due to the momentum transfer rate be more significant at high flow rate. More experimental observations shall be made, including in those planes placed above that one observed during the present work realization.

## 4. Conclusion

In this work an evaluation process of a facility which was designed to be used for jet induced flow investigation is shown. Metrological requirements and concepts guided all steps of the experimental work, from the mounting facility up to the limits of the bench use determination. A StereoPIV system was employed to identify the resulting flow behavior of the axisymmetric jet in the stagnant environment. The results indicated that with a nozzle of inner diameter 1mm the best range of flow rate for working in the bench with the bypass totally closed goes from 0.05 to 0.2 L/min. This is the total flow rate which pass through the nozzle. It is important to mention that this range can be different if the nozzle diameter or the number of nozzles change, since it can modify the pressure conditions inside the pipeline. Depending on the situation the bypass aperture can be demanded. Future works in this bench will include experiments using different configurations of the nozzles.

As additional result of this work, an experimental procedure for PIV use at the Fluid Dynamics Metrology Division (Dinam/Inmetro) was improved, thus contributing in sense of metrological approach considerably. It was extremely important for the Dinam/Inmetro researchers knowledge, since the uncertainty analysis for imaging of fluid flow is a research line which is being introduced as essential for the activities in the Division. In this work, the uncertainties analyses related to the flow field mapping are not presented, since the issue is complex and it is being investigated. However, during the work, in several steps of using the imaging acquisition system, errors were quantified and minimized.

The experiments here described had the purpose of confirming the applicability of the bench for studies on the hydrodynamics of mixing process induced by injection of liquid in liquid. In the environmental research area, this theme is fundamental for understanding pollutant dispersion mechanisms.

The bench was approved to be used according to the expected, and from now it will be a facility for supporting the researches on fluid dynamics investigations at Inmetro.

## Acknowledgments

The authors are grateful to CNPq, FAPERJ, FINEP, PETROBRAS, PFRH-PETROBRAS and INMETRO for their support and sponsorship which have become possible the development of this research.

## References

[1] Lee W H and Chu V H 2003 *Turbulent jets and plumes - a Lagrangian approach* (Norwell: Kluver Academic Publishers)
[2] Carazzo G, Kaminski E and Tait S 2006 *J. Fluid. Mec.* **547** 137
[3] Simiano M and Lakehal D 2012 *Int. J. Mult. Flow.* **47** 141
[4] Adrian R J and Westerweel J 2011 *Particle image velocimetry.* **4th** Ed. (New York: Cambridge University Press)
[5] Goldstein R J 1996 *Fluid mechanics measurement.* **2nd** Ed. (Washington: Taylor and Francis)
[6] Borg A, Bolinder J and Fuchs L 2001 *Exp. Fluids.* **31** 140
[7] Farias M H and Santos A M 2014 *Abst. book - 8th Int. Conf. Adv. Comp. Eng. Exp.* **art.** 242

# Study for the requalification of Inmetro's Primary Hardness Standardization Machine for Vickers HV3 scale

**J T Filho[1,2], R Q Fratari[2], C R S Azeredo[2] and S P Oliveira[2]**

[2] Force Laboratory, National Institute of Metrology Quality and Technology, Av. Nossa Senhora das Gracas, 50, Predio 3, Duque de Caxias, RJ 25250-020, Brazil

[1] E-mail: jtfilho@inmetro.gov.br

**Abstract.** The standardization of Vickers hardness quantity in Brazil was established as related to the quality management system and the properties of metallic materials that are used in industries like steelworks, car makers aircraft manufacturers. This work aimed in describing the metrological methodology applied in the requalification of the Inmetro´s Primary Hardness Standardization Machine for low, medium and high hardness ranges of Vickers HV3 scale.

## 1. Introduction

Hardness is the most used mechanical property and has widespread recognition due to its essential importance in both production and service performance of many items produced in virtually all industries, the most important being the metal-mechanical, steel, metallurgy, automobile, aerospace, tools and machinery and equipment sectors. There is no unique definition for hardness. It can be defined as the resistance to permanent plastic deformation, to absorbed energy during the impact, and to the risk or penetration of a harder material in a softer one. Industrially, the last approach is the one used for metals.

Hardness measurements are run several times in the production line in order to assure the quality of various items produced by sampling some parts from time to time along the production lines and in raw-materials receiving tests. It is the most evaluated property during the research and development (R&D) activities made in universities, R&D centers and industries in the metal-mechanical sector. The hardness results are also used in the design and development of new materials, products, processes, methods and innovative technologies. Among the existing indentation hardness scales, the most used are Brinell, Vickers and Rockwell ones [1].

The Hardness quantity is also the subject of standardization actions by standardizing bodies which issue international standards (ISO, etc.), regional ones (Mercosul NM, etc.) or national ones – Brazilian standards (ABNT- NBR), American standards (ASTM, etc.) – as well as through the Guide of calibration in hardness, which establishes best practices for obtaining the measurement uncertainty in hardness tests, issued by EURAMET [2], body that consolidates all NMIs (National Metrology Institutes) in Europe.

[1] To whom any correspondence should be addressed.

The last paragraph emphasizes the importance of the reliability of the measurement results, reported in metrology as metrological reliability. This feature implies that each measurement shall be related to stated references, or have what is called traceability to national or international standards through an unbroken chain of comparisons each of them having, due to the inherent randomness (variability) of any measurement (since there are several causes for this), some estimates of the involved measurement uncertainties.

To ensure that the metrological reliability is maintained over time, Inmetro's Primary Hardness Standardization Machine has been its main metrological parameters controlled by Shewhart control charts. In case it is necessary any change in the system, no matter the type of event – for example, improvement in terms of accuracy/precision or maintenance of the standardization system –, a system requalification will have to be made.

The objective of this work is to show the procedure for requalification of the primary machine focusing on the application of force during the standardization process of Vickers HV3 hardness scale. The tools used in the requalification process will be described and their importance to keep the standardization system with a suitable metrological quality will be stressed as well. Therefore, if there were no gaps in the traceability chain, the hardness measurements obtained in any durometer (e.g. a hardness testing machine) installed in any industry will provide reliable results. Moreover, they will be considered as traceable to the primary hardness standard of a country (in Brazil, the hardness national standard is Inmetro's Primary Hardness Standardization Machine) and, indirectly, to international standards by international comparisons between Inmetro and other NMIs – under the auspices of the BIPM (International Bureau of Weights and Measures), the Inter-American Metrology System (SIM) or South American Metrology (SURAMET).

## 2. Materials and Methods

The metrological standards used here were the Primary Hardness Standardization Machine (PHSM) – where there is the realization of the quantity hardness by the way of the standardization of Brinell, Rockwell and Vickers hardness scales (figure 1A) – and the Metrological Reference System for Brinell and Vickers Hardness Scales (Gal-Vision), figure 1B. Relating to the latter new image patterns to enable the determination of the Vickers impression diagonal were developed in the image analysis system, so that the hardness measurements and calibrations could be performed with the highest metrological quality.

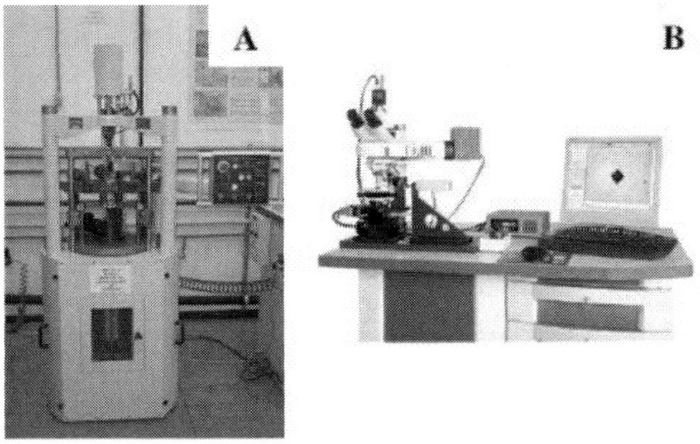


**Figure 1.** Inmetro's Primary system of Rockwell, Brinell and Vickers hardness scales: (A) Primary Hardness Standardization Machine (PHSM); (B) Metrological Reference System for the Brinell and Vickers Hardness Scales (Gal-Vision).

Objective lenses of magnifications/numerical apertures 2.5x/0.07, 5x/0.12, 10x/0.25, along with a 20x/0.40 lens for adjustment of the Vickers indentation, were used sequentially in the microscope of the Gal-Vision. This procedure was necessary for the measurement of the Vickers diagonals. The Vickers hardness was calculated according to equation 1 [2]:

$$HV = 0.102 \cdot \frac{2F \cdot sen68^{o}}{d^{2}} \quad (1)$$

where:
HV – value of Vickers hardness;
F – testing force during the standardization, in newtons;
d – average length of the diagonals of the indentation, in millimeters.

A 100 N class 00 force transducer was calibrated according to standard ISO 376 [3] was used to check out the applied force during the standardization process. The HV3 hardness values were analyzed before and after the requalification process in such a way it was possible to assure the metrological reliability of both PHSM and Gal-Vision.

International hardness standards ISO 6507-1:2005 [4], 6507-2:2005 [5] and 6507-3:2005 [6] were applied in this study research, which provided the most important parameters of the measurement process that directly influence the Vickers hardness calibration. For comparing before and after requalification´s calculations with data coming from hardness measurements, relevant methodologies described in ISO GUM [7] and EURAMET Guide [2] were used. Equations 2 and 3 represent the metrological essence of this requalification, since they are the basis for calculating both the uncertainty regarding the repeatability and the normalized error:

$$u_{rep} = t_{student} \cdot \frac{\sqrt{\frac{\sum\left(\overline{X} - X_{i}\right)^{2}}{n-1}}}{\sqrt{n}} \quad (2)$$

where:
$u_{rep}$ – standard uncertainty of measurement due to repeatability;
$t_{student}$ - sample size correction factor;
n – number of measurements;
$X_i$ – any measurement i;
$\overline{X}$ – arithmetic mean of measurements.

$$E_{n} = \frac{\overline{X_{i}} - \overline{X_{ref}}}{\sqrt{U_{i}^{2} + U_{ref}^{2}}} \quad (3)$$

where:
$E_n$ – normalized error;
$\overline{X_i}$ - mean hardness HV3 value after the adjustment of the force;
$\overline{X_{ref}}$ - mean hardness HV3 value before the adjustment of the force;
$U_i$ - expanded measurement uncertainty after the adjustment of the force;
$U_{ref}$ - expanded measurement uncertainty before the adjustment of the force.

## 3. Results and Discussion

The results presented below are connected to the PHSM requalification process for the Vickers hardness scale. The scale chosen for the study was the HV3 one, whose scale uses a 3 kgf force. This force is applied to the indenter, which is pressed against the surface of a reference hardness block. This choice was based on the lower force limit of the PHSM, never developed and put in operation before which means the use of the smallest force (3 kgf) to perform standardization in hardness metrology in Brazil. Thus, the application of force is one of the parameters that need to be monitored because of its high sensitivity to changes in any part of the system. Figure 2 shows a picture of the existing force systems that operate on the PHSM in the standardization process of HV3 hardness scale.

**Figure 2.** Force systems that operate during the standardization of HV3 hardness scale: force compensation and force application ones.

Figure 2 shows that the force application system in case of HV3 is connected to the force compensation system. The application of force of the PHSM is performed by a deadweight system, in other words, gravity acceleration acts directly on the set of stainless steel masses of the system. For HV3 scale, only the framework of the application of forces is acting and its value is equivalent to a 10 kgf selection of masses in the force application system. Thus, to reach a value of 3 kgf it´s necessary the use of a force compensation system which, in this case, operates with a load of 7 kgf in the opposite direction to the force application system, resulting in a net force of 3 kgf necessary to perform the HV3 scale.

According to ISO 6507-3:2005, in performing the hardness standardization a machine has to present an absolute percentage error (Error%) less than 0.10% [6]. Thus, adjustments may be necessary in the mass used to generate the force, so that the Error% is in accordance with the limits established by the pertinent standard. Table 1 shows the results for the measurement of PHSM system relative to the application of a 3 kgf nominal force.

**Table 1.** Force verification of the PHSM for a 3 kgf nominal force.

| Nominal Force/N | Force/N | Force/kgf | Error/kgf | Error%/% |
|---|---|---|---|---|
| 29.362 | 29.25726 | 2.989252 | -10.7482 | -0.358 |
| 29.362 | 29.25776 | 2.989303 | -10.6972 | -0.357 |
| 29.362 | 29.25876 | 2.989405 | -10.5951 | -0.353 |
| 29.362 | 29.25976 | 2.989507 | -10.4931 | -0.350 |
| 29.362 | 29.25926 | 2.989456 | -10.5441 | -0.351 |

It can be observed in table 1that the Error% values were out of the ± 0.10% limits allowed by the ISO 6507-3:2005 [6]. In fact, there is a case of negative Error% i.e. it means there is a lack of masses to attain the nominal force value. So, it was concluded that the best way to increase the metrological quality of the force application process for the standardization of HV3 scale was the addition of specific amounts of tiny masses to the force application system of PHSM. A detailed study, comprising five different conditions C1 to C5, was conducted by the addition of mass increments to the PHSM as shown in Table 2.

**Table 2.** Study of the force applied in the PHSM for a 3 kgf nominal force by using tiny masses adjustment.

| Condition | Nominal Force/kgf | Adjustment/g | Force/kgf | Error%/% |
|---|---|---|---|---|
| C1 | 3 | 0 | 2.9894 | -0.354 |
| C2 | 3 | 9 | 2.9985 | -0.049 |
| C3 | 3 | 10 | 2.9996 | -0.015 |
| C4 | 3 | 11 | 3.0006 | 0.019 |
| C5 | 3 | 15 | 3.0045 | 0.150 |

In table 2, the reported values of nominal forces and Error% were obtained from an average of five measurements. Accordingly, it can be noted that condition C3 presented the minor Error% value, being considered the most appropriate for the requalification process of the HV3 hardness scale at PHSM. Figure 3 shows graphically the effect of the addition of masses to the PHSM on the Error% related to the force applied during the standardization of HV3 scale.

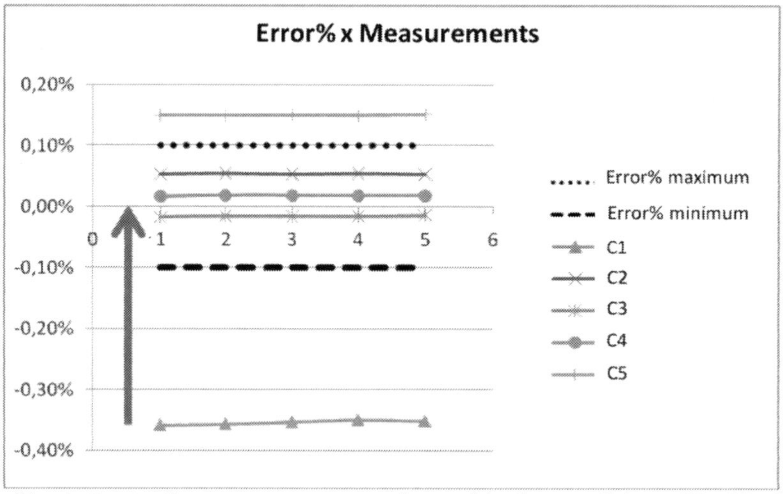


**Figure 3.** Error% relative to the measurements performed in the conditions C1, C2, C3, C4 and C5.

Figure 3 is a corresponding graph of Table 2. It can be noticed that the mass increments promotes a trend that is represented by the vertical red arrow, in the direction to the decrease of the absolute value of the Error%. So, through the analysis of figure 3, it can be observed that the condition C3 is the one that has the lowest absolute value for the Error%. Thus, during the standardization made to HV3 scale at PHSM it was necessary the addition of a 10 g mass to the force application system.

On finishing the above analysis to determine the best condition for the application of force in the PHSM, the first part of the task of requalifying study was realized. The next and last step was a verification of the performance of PHSM related to standardization process itself under this new operating condition. It means reference hardness blocks calibrated before the requalification were recalibrated and their results compared to low, medium and high hardness ranges of HV3 scale.

Table 3 presents the average values of hardness, standard uncertainty of measurement due to repeatability and expanded measurement uncertainty to the low, medium and high hardness ranges of HV3 scale before and after the requalification of PHSM.

**Table 3.** Performance study of PHSM during the standardization of HV3 scale.

| Parameters | HV3 low hardness | | HV3 medium hardness | | HV3 high hardness | |
|---|---|---|---|---|---|---|
| | Before | After | Before | After | Before | After |
| Average hardness | 131.7 | 131.0 | 568.2 | 565.7 | 848.2 | 844.9 |
| Standard uncertainty of measurement due to repeatability | 0.8 | 0.8 | 3.1 | 1.2 | 2.5 | 1.4 |
| Expanded measurement uncertainty | 2.5 | 2.6 | 6.4 | 3.0 | 5.3 | 3.4 |

In table 3 it was found that for low HV3 hardness, the force requalification process did not cause any significant changes in hardness since the hardness indentation size is big, so its diagonals were easily read. As a consequence, the related expanded measurement uncertainties, before and after the HV3 requalification, had nearly no changes. However, for medium and high ranges of HV3 hardness - that have smaller indentation sizes - it can be said the requalification process was effective and improved the metrological quality of the measurements, since it was possible to obtain a reduction in the expanded uncertainty. Evidence of this comes from the decrease of 53% (from 6.4 to 3.0 HV3) for the medium HV3 hardness range and 36% (from 5.3 to 3.4 HV3) for the high HV3 one.

Figure 4 shows the review of the normalized errors with respect to standardization of HV3 scale for low, medium and high hardness ranges before and after the force adjustment process.

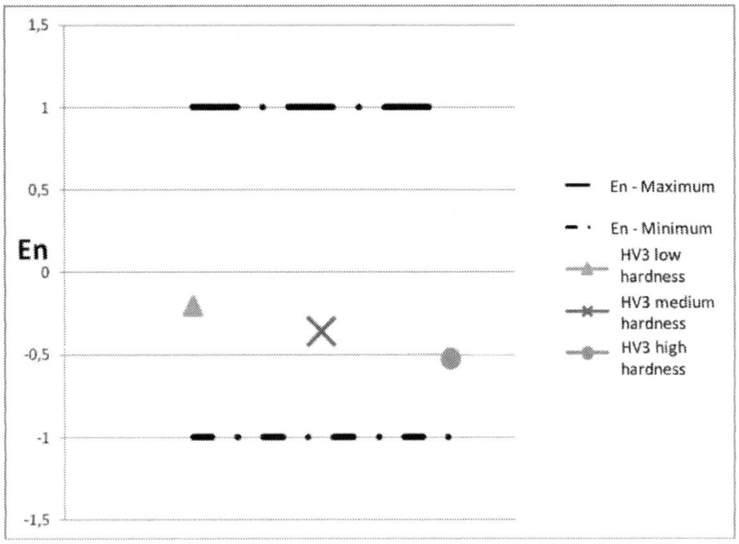


**Figure 4.** Normalized errors for low, medium and high HV3 hardness ranges.

Figure 4 depicts all calculated normalized errors taking into account the obtained results before and after the force requalification process: they are within the normalized error limits of 1 and -1. In other words, this result showed that not only the requalification promoted an improvement in the metrological quality of PHSM but also its standardizing measurements are still reliable. In metrological terms, these results indicate that both the PHSM and the Gal-Vision metrological standards have the necessary prerequisites for the HV3 scale standardization, since there was a high similarity between the normalized errors before and after the force adjustment for lower, medium and high hardness ranges. Then, the methodology applied in this work can eventually be applicable to others hardness scales in their requalification processes.

## 4. Conclusions

(i) This work showed the metrological methodology applied in the requalification of the Inmetro´s Primary Hardness Standardization Machine for low, medium and high hardness ranges of HV3 scale.

(ii) Results in terms of normalized errors obtained with the requalification of the PHSM for HV3 Vickers hardness scale were satisfactory, since they were within the limits of 1 and -1.

(iii) Inmetro´s PHSM is suitable for performing the standardization of HV3 scale in Brazil, since the % relative error to the nominal applied force was less than 0.10%.

(iv) On comparing the expanded uncertainty of the HV3 scale before and after requalification it yields an increase in the metrological quality of the results for middle and high HV3 hardness ranges because there was a reduction in the uncertainty of results for both hardness ranges.

(v) The normalized error, as a result of a comparison of measurements reliability before and after the requalification process, has shown a high similarity between them. This was an evidence that the parameters involved in the standardization process remain under control at the end of the requalification process.

## 5. Acknowlegment

The authors thank to FAPERJ (Research and Development Support Foundation of Rio de Janeiro State), for the research support grant for the technologist Celso Ricardo da Silva Azeredo, and also to Inmetro (National Institute of Metrology, Standardization and Industrial Quality) where all of the researchers of this study work in the Force Laboratory (which deals with areas of force, torque, hardness and impact), for supporting the development of this work.

## 6. References

[1] William D. Callister, Jr. "Materials Science and Engineering – An Introduction". New York/USA: John Wiley & Sons, Inc., 7th Edition, 2007.

[2] "Calibration Guide EURAMET/cg-16/v.01: Guidelines on the Estimation of Uncertainty in Hardness Measurements". Braunschweig/Germany: European Association of National Metrology Institutes, 2007.

[3] ISO 376:2011 "Metallic materials – Calibration of Force-proving Instruments Used for the Verification of Uniaxial Testing Machines", Geneva/Switzerland: International Organization for Standardization, 2011.

[4] ISO 6507-1:2005 "Metallic materials – Vickers hardness test – Part 1: Test method". Geneva/Switzerland: International Organization for Standardization, 2005.

[5] ISO 6507-2:2005 "Metallic materials – Vickers hardness test – Part 2: Verification and Calibration of Testing Machines". Geneva/Switzerland: International Organization for Standardization, 2005.

[6] ISO 6507-3:2005 "Metallic materials – Vickers hardness test – Part 3: Calibration of Reference Blocks". Geneva/Switzerland: International Organization for Standardization, 2005.

[7] JCGM Member Organizations (BIPM, IEC, IFCC, ILAC, ISO, IUPAC, IUPAP and OIML). "JCGM 100:2008 – GUM 1995 with minor corrections – Evaluation of Measurement Data – Guide to the Expression of Uncertainty in Measurement". Sèvres/France: Joint Committee for Guides in Metrology/Bureau International des Poids et Mesures, 2008.

Journal of Physics: Conference Series **648** (2015) 012024 doi:10.1088/1742-6596/648/1/012024

# Systematics for checking geometric errors in CNC lathes

**R P Araújo[1] and T L Rolim [2]**

1 Rogério Pontes de Araújo, Recife, Brazil
2 Tiago Leite Rolim, Recife, Brazil

E-mail: rogeriopontes10@hotmail.com

**Abstract:** Non-idealities presented in machine tools compromise directly both the geometry and the dimensions of machined parts, generating distortions in the project. Given the competitive scenario among different companies, it is necessary to have knowledge of the geometric behavior of these machines in order to be able to establish their processing capability, avoiding waste of time and materials as well as satisfying customer requirements. But despite the fact that geometric tests are important and necessary to clarify the use of the machine correctly, therefore preventing future damage, most users do not apply such tests on their machines for lack of knowledge or lack of proper motivation, basically due to two factors: long period of time and high costs of testing. This work proposes a systematics for checking straightness and perpendicularity errors in CNC lathes demanding little time and cost with high metrological reliability, to be used on factory floors of small and medium-size businesses to ensure the quality of its products and make them competitive.

## 1. Introduction

Machine tools play a significant role in the development of a country, and are essential for modern metal-mechanical industry and in evolution of machinery and equipment, adding improvements to human quality of life by manufacturing consumer goods that provide comfort and well-being. Machine tools have undergone significant evolution in the last 30 years with the contribution from various fields of knowledge (mechanics, electronics, information technology and new materials) [1], which evolved from simple mechanical machines to automated machines with mechatronic systems, through the introduction of the computer numerical control (CNC). In the 80's, about 10% of the machines produced in the most highly industrialized nations were machine tools, and approximately 10% of the workforce were focused on building these machines [2]. Under the current scenario, companies seek more and more to adopt quality standards in order to gain competitiveness, and thus seek to implant in their production processes speed, flexibility, high performance, reliability and cost reduction, because if they do not comply to these standards they will quickly reach failure [3]. However, it is not easy to achieve excellence in the face of numerous difficulties that arise on a daily basis, it is necessary to look continuously for improvements in all steps involved in the production process [4]. Common are the losses generated by a manufacturing system due to its inefficiency, which in turn raises the costs of the process due to waste of time and materials. It is not different for machining processes, which have great loss of material but give the piece a high added value, since they are commonly used as a final step in the manufacturing process, providing the shape, size and finish of the material. However, accuracy levels can be achieved in the machining process which are impractical in other processes, justifying its use especially in finishing processes [5-6].

Machine tools are not perfect and have imperfections that reflect directly on the geometry and dimensions of machined parts. In order to minimize those undesirable effects, improvements were implanted initially in the mechanical assembly, but mechanical optimization alone will not eliminate the geometric and dimensional errors from the non-idealities in the mechanical assembly of the machines. Geometric testing of machine tools is used alternatively. Besides providing information on the possible deviations of shape and size generated by the machine tool, the user finds out its dimensional tolerance in manufacturing parts [7]. The geometrical errors are present in all machine tools, primarily as a consequence of mechanical imperfections and misalignment of machine elements. These errors cause wrong positioning and orientation of the cutting tool relative to the workpiece. For each axis of a machine tool there are 6 sources of geometric errors, three of translation, where one of these errors relates to position and the other two are of straightness, and three angular errors known as *roll*, *pitch* and *yaw*. Also, there are perpendicularity errors between the axes. Figure 1 shows the movement scheme of a machine guide indicating the error components present. Considering a machine tool that has three linear axes perpendicular to each other, there will be 21 sources of geometric errors.

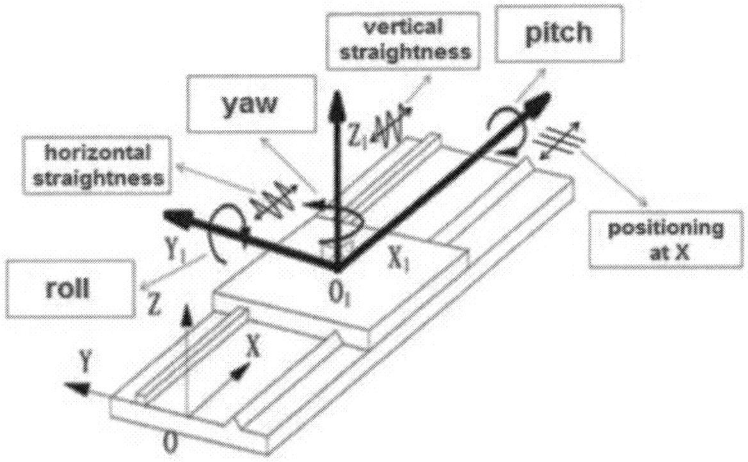


**Figure 1.** Geometric errors with movement along axis X (adapted from OKAFOR, 2000).

## 2. Objective

The aim of this work is to propose a systematics for checking straightness and perpendicularity errors demanding little time and low cost with high-reliability metrology to be applied in CNC lathes, which can be used on the factory floor by users of small and medium-sized companies to thereby assess the geometric errors of their machines and ensure the quality of their end product so that they can remain competitive because, despite the importance of the use of geometric tests as a tool aiding in quality control, they are not yet applied on a large scale due to the fact that traditional methods are expensive and demand long periods of time.

## 3. Materials and methods

Steel billets SAE 1020, 349 mm long and with a diameter of 58 mm, were used, following preset cutting parameters, and three specimens (staggered cylindrical - Figure 2, and non-staggered cylindrical - Figure 3) were machined, using an almost new Romi CNC Centur 30D lathe (Figure 4).

The cutting parameters were determined carefully in order to minimize maximum efforts involved in the machining process and thereby avoid bending deformation of the tool-holder and specimen. The values used were: cutting depth 0.5 mm, cutting speed 200 m/min, roughing operation advancement 0.3 mm/rev and advancement in finish operation of 0.2 mm/rev.

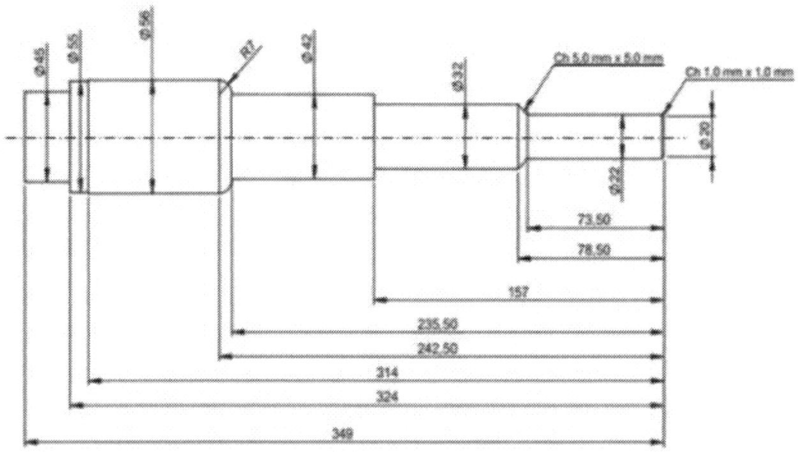

**Figure 2.** Staggered cylindrical piece (in mm).

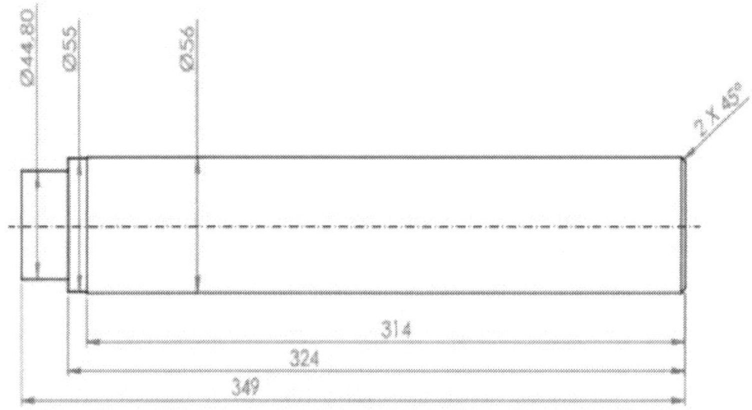


**Figure 3.** Non-staggered cylindrical piece (in mm).

**Figure 4.** CNC ROMI Centur 30D Lathe.

For each workpiece, a brand-new cutting edge was employed, and cutting tools used were made of hard metal for turning, Sandvik Coromant, ISO code CMT 11 T3 08-UR 4025. A 25 mm x 25 mm square-section Sandvik Coromant tool-holder was used (ISO Code SDJCR 2525M 11) (Figure 5).

**Figure 5.** Tool-holder for turning.

Six specimens were machined: three staggered pieces and one non-staggered piece by using MQL (compressed air and cutting oil) and three staggered pieces and one non-staggered piece using cutting fluid in abundance, to determine which of the two conditions has a better performance.

3.1. Thermography of the specimens
Through a FLIR SYSTEMS (FLIR SC 660 model) thermographic camera (Figure 6), thermographic filming was performed (Figure 7) on the piece machined on the two lubrication conditions, in order to check temperature distribution along the parts.

**Figure 6.** FLIR FC 660 Thermographic Camera.

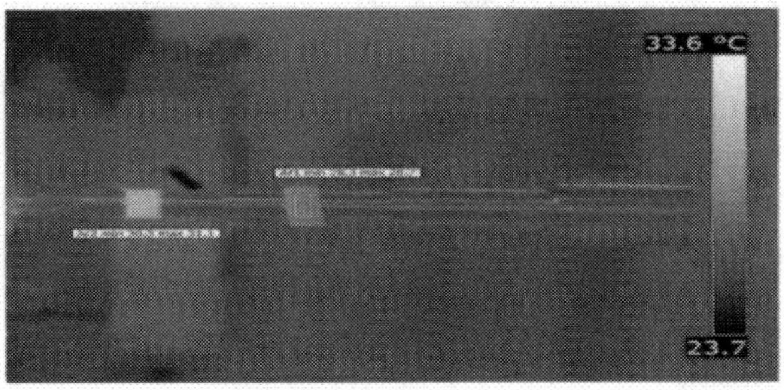


**Figure 7.** Thermography of a machined piece.

After thermographic analysis, the maximum temperature values obtained were (75.1 ± 1.5) °C using MQL and (31.1 ± 1.5) °C using abundant cutting fluid.

3.2. Scanning Electron Microscopy of cutting tools after machining
Through the HITACHI TM 3000 scanning electron microscope (Figure 8), images were obtained (60x) of both the brand-new and the used cutting tools after machining operation (Figures 9 and 10), with both lubrication conditions, in order to perform a qualitative analysis of possible cutting tool wear and breakdown.

Journal of Physics: Conference Series **648** (2015) 012024 doi:10.1088/1742-6596/648/1/012024

**Figure 8.** HITACHI TM 3000 scanning electron microscope.

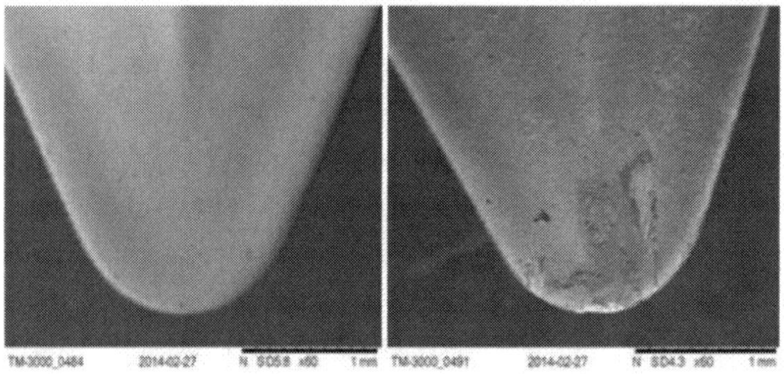

*(a)* *(b)*

**Figure 9.** Comparison of the new cutting tool (a) to the used one with MQL (b).

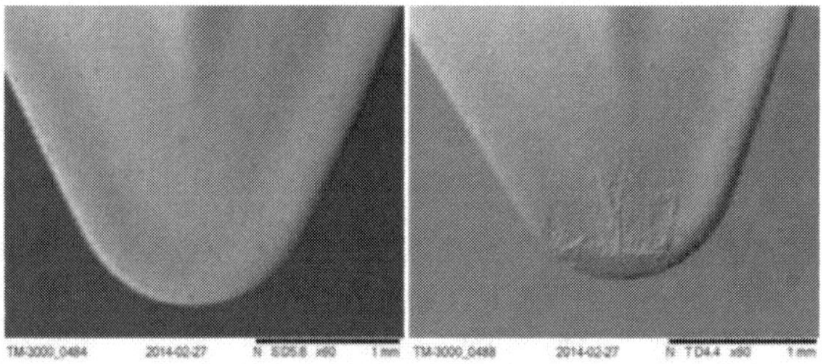

*(a)* *(b)*

**Figure 10.** Comparison of the brand-new cutting tool (a) to the used one with abundant cutting fluid (b).

3.3. Cutting force calculation according to Kienzle

In 1951 , Kienzle created a quite accurate formula the result of which is closest to the actual values observed in practice through direct measurement of machining forces.

For this reason this formula is the most widespread and used today [8]:

$$F_c = K_{s1} \, . h^{1-z} . \, b \qquad (1)$$

$K_{s1}$ and $z$ are the constants of the material, $h$ is the nominal thickness of cut and $b$ is the chip length.

**Table 1.** Kienzle's coefficient values for SAE 1020 steel.

| Material | $K_{S1}$ *(N/mm²)* | *1-Z* |
|---|---|---|
| SAE 1020 steel | 1800 | 0,83 |

Replacing the Kienzle coefficient values for SAE 1020 steel in equation (1) gives the following cutting values: 331.40 N for the roughing condition and 236.70 N for the finishing condition, which in turn are too small values to cause bending in tool-holder and/or in the part.

### 3.4. Computer Simulation

Through the ANSYS 14.0 software computer simulations were performed considering the cutting strength values found by using equation (1) in order to check tool-holder behavior according to the loads applied in both the roughing and finishing conditions. Figure 11 shows the result of computer processing.

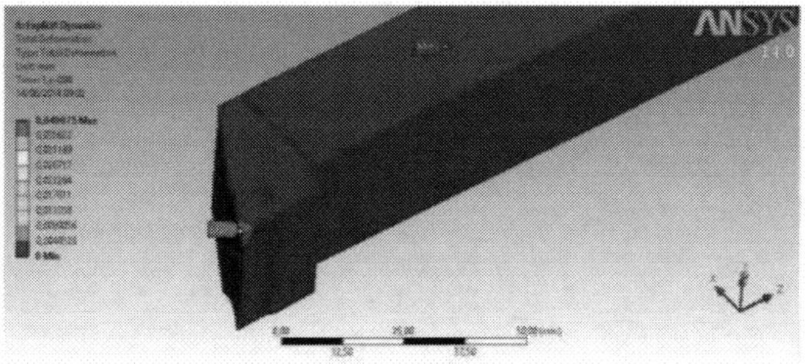


**Figure 11.** Computer tool-holder simulation.

### 3.5. Roughness measurement of non-staggered piece

Using an Instrutherm RP-200 rugosimeter, shown in Figure 12, measurements were performed on several parts of the non-staggered piece, resulting in the average roughness value of (1.032 ± 0.018) μm.

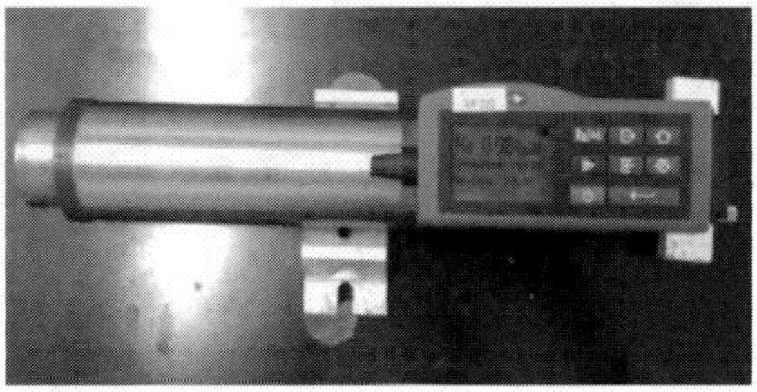

**Figure 12.** Measurement of roughness on the piece.

### 3.6. Measurement of pieces

After the machining operation, straightness measurements were performed on the non-staggered pieces and of perpendicularity on the staggered pieces with a Mitutoyo Beyond - Crysta C measuring machine by coordinates, shown in Figure 13, with calibration certificate number (RBC) 03206/13.

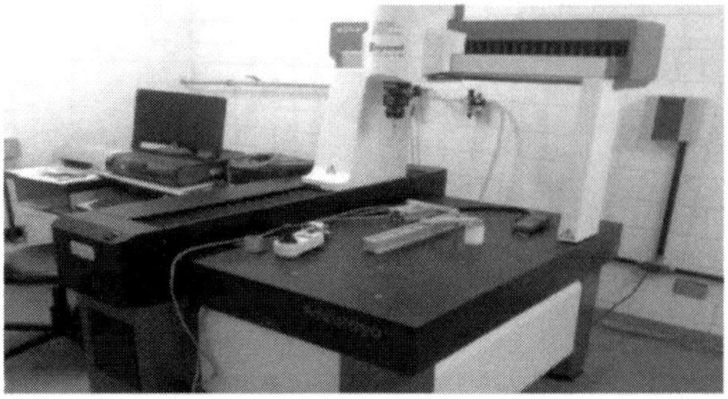

**Figure 13.** Measurement of pieces.

Measurements of straightness and perpendicularity of pieces were carried out at a temperature of (20 ± 1) °C, whose control was performed with a MINIPA MTH model 1380 thermo-hygrometer, shown in Figure 14, with calibration certificate No. 6285/14 (REMESP).

**Figure 14.** MINIPA MTH-1380 thermo-hygrometer.

3.7. Measurement results

Table 2 shows the mean values of performed measurements: straightness for non-staggered pieces and perpendicularity for staggered pieces.

**Table 2.** Straightness and perpendicularity values.

| | | | | |
|---|---|---|---|---|
| Straightness | (6,48 ± 2,82) μm | | | |
| | Length | Piece 1 | Piece 2 | Piece 3 |
| Perpendicularity on plane x-y, μm | $Ø_1$ = 56 mm | 10,42 ± 4,05 | 13,57 ± 4,20 | 10,42 ± 4,05 |
| | $Ø_2$ = 42 mm | 10,18 ± 4,01 | 13,57 ± 4,20 | 10,66 ± 4,09 |
| | $Ø_3$ = 32 mm | 10,66 ± 4,06 | 13,33 ± 4,18 | 11,39 ± 4,12 |
| | $Ø_4$ = 22 mm | 9,66 ± 3,81 | 12,84 ± 4,15 | 11,63 ± 4,14 |

.

## 4. Results and discussion

From the analysis of the results it was evident that the machining condition using cutting fluid in abundance should be used for verification of errors in the proposed system. This finding can be observed by the state of the tool which presented to the electronic microscope a light wear without compromising the geometry of the workpiece. From the calculated values of cutting force it was found by the classical equation of Kienzle and computer simulation that the tool holder did not deform during machining operations, thus not compromising the geometry of the workpiece.

With the roughness measurement of the machined specimens, it was concluded that the specimens had good surface finish and average roughness values well below the straightness values.

## 5. Conclusions

The systematics presented here provides machining operations with minimal effort, low temperatures and good surface finish of the parts, with little wear of the cutting tools, allowing a good quality of form and dimension of the pieces obtained. Therefore, it can be concluded that it is a simple and low-cost systematics, with high practicality of application for checking straightness and perpendicularity errors on lathes with computerize numerical control with a reduced bus. This checking systematics is presented as an essential tool and an option for decision making when choosing more sophisticated testing methods.

## 6. References

[1] Rolim T L 2003 *Sistemática Indicadora de Método para Calibração de Máquinas de Medição por Coordenadas* Doctoral thesis in mechanical engineering (Brazil: Federal University of Paraíba)

[2] Weck M 1984 *Handbook of Machine Tools – Metrological analysis and performance tests* vol 4 John Wiley and Sons

[3] Souza, André Roberto 2000 *Padrões Corporificados e Tecnologia de Medição por Coordendas – Inovando a Qualificação Geométrica de Centros de Usinagem* Doctoral thesis in mechanical engineering (Brazil: Federal University of Santa Catarina)

[4] Ueno, Shigeru 1996 *A máquina da próxima geração tem múltiplas funções e é rápida e precisa* Máquinas & Metais pp 42-53

[5] Ferraresi D 1977 *Fundamentos da usinagem dos metais* vol 1 ed Edgard Blücher (Brazil: São Paulo).

[6] Grasson Thomas 1997 *Machine tool builder's future goes on display* American Machinist pp 154 – 167

[7] Eckhardt M et al. 2008 *Avaliação geométrica de máquinas-ferramenta CNC através do método de padrões corporificados* Máquinas & Metais pp 120 – 129

[8] Erdel Bert 1997 *O impacto da usinagem a alta velocidade na produção* Máquinas & Metais pp 36 – 41

# Analysis of drift correction in different simulated weighing schemes

**A Beatrici[1], A Rebelo[2], D Quintão[3], F L Cacais[4], V M Loayza[5]**

[1, 2, 3, 4, 5] Mass Laboratory – Lamas, Instituto Nacional de Metrologia, Qualidade e Tecnologia – Inmetro – Brazil

E-mail: [1]abeatrici@inmetro.gov.br , [2] asjunior@inmetro.gov.br , [3]dnquintao@inmetro.gov.br , [4] facacais@inmetro.gov.br , [5]vmloayza@inmetro.gov.br

**Abstract**. In the calibration of high accuracy mass standards some weighing schemes are used to reduce or eliminate the zero drift effects in mass comparators. There are different sources for the drift and different methods for its treatment. By using numerical methods, drift functions were simulated and a random term was included in each function. The comparison between the results obtained from ABABAB and ABBA weighing series was carried out. The results show a better efficacy of ABABAB method for drift with smooth variation and small randomness.

## 1. Introduction

When the accuracy limit of a weighing equipment is reached, for example in highest accuracy mass standards calibration, mass comparators (high resolution short range electronic balances) often exhibit drift of zero (nonzero indication when there is no load on the weighing plate) which adds up to weighing indications. This drift is caused by influence of several factors such as self-heating and environmental conditions changes. This *zero drift* can by far outweigh the comparator´s last significant digit. In an ultramicro mass comparator with 0.1 microgram resolution may add up to tens of micrograms in balance indication during a calibration weighing series taking typically about 20 to 30 minutes long. Thus, drift would become an important source of uncertainty if not properly treated.

Calibration of high accuracy mass standards is based on the determination of the mass difference between a test standard (object) and a reference standard. In the zero drift treatment comparative weighing cycles are widely used, as an example we can mention the ABA and ABBA (A = reference standard weighing, B = test standard weighing). These weighing schemes assume that within a relatively short time interval the drift has a linear shape, and the average of the differences between the final and initial weighings within each cycle realized at equal time intervals (or symmetrical), can eliminate the zero drift in the balance indicator. Experimentally the drift shows numerous behaviors and are best defined as a sum of some functions, but in a limited period of time we can recognize linear, power, or periodic (i.e. sine, cosine) behavior and even of random nature.

In this work we apply simulated drifts using functions of some nature: linear, quadratic, cubic, sine and to all of them it was added a random term in order to compare their performance (effectiveness) in the reduction or elimination of the drift effects arising from weighing equipment. We also present results from real measurements obtained by applying the weighing schemes in weighing series using the mass comparator Mettler-Toledo AT1006.

[1] Address: Av. Nossa Senhora das Graças, 50. Xerém - Duque de Caxias – RJ. ZIP: 25250-020

Journal of Physics: Conference Series **648** (2015) 012025 doi:10.1088/1742-6596/648/1/012025

## 2. Drift in weighing cycles

In order to determine the conventional mass of standard weights [1], calibration laboratories perform comparative weighing cycles of test standard against a reference standard [2]. Such weighing cycles are intended to reduce or eliminate errors due to systematic effects not correctable, usually environmental, which act during individual weighings of the standards. These errors can be considered as an additive term in the mathematical model for the indication of the scale $I$ [3] according to equation (1).

$$I = k \cdot g \cdot \left\{ m_c \cdot \left[ \frac{\left(1-\frac{\rho_0}{\rho_c}\right)\cdot\left(1-\frac{\rho_a}{\rho_{(T)}}\right)}{\left(1-\frac{\rho_0}{\rho_{(20°C)}}\right)} \right] \right\} + \delta_T + \delta_M + \delta_E + \delta_S + \delta_X + \delta_R + \delta_N + \delta_A + f(t) \qquad (1)$$

Where:

$k$ adjust constant;
$g$ local gravity acceleration;
$m_c$ conventional mass of the standard weight;
$\rho_0$ conventional air density (1.2 kg/m$^3$);
$\rho_c$ conventional density of the standard weight (8000 kg/m$^3$);
$\rho_a$ air density during weighing;
$\rho_{(T)}$ density of standard weight, at temperature $T$ of the standard weight;
$\rho_{(20°C)}$ density of standard weight at reference temperature 20 °C;
$\delta_T$ error due to thermal effects;
$\delta_M$ error due to magnetic effects;
$\delta_E$ error due to electrical effects;
$\delta_S$ error due to surface effects;
$\delta_X$ eccentricity error due to the standard weight on the weighing plate;
$\delta_R$ resolution error of balance indication;
$\delta_N$ nonlinearity error of the balance;
$\delta_A$ random error;
$f(t)$ zero drift of balance indication at weighing time $t$.

The mathematical model for the indication difference $\Delta I$ of weighing of a reference standard and a test standard with similar densities, under stable environmental conditions, can be determined from equation (2).

$$\Delta I = k \cdot g \cdot \left\{ (m_{cob} - m_{cp}) \cdot \left[ \frac{\left(1-\frac{\rho_0}{\rho_c}\right)\cdot\left(1-\frac{\rho_a}{\rho_{(T)}}\right)}{\left(1-\frac{\rho_0}{\rho_{(20°C)}}\right)} \right] \right\} + \varepsilon_{T,M,E,S,X,R,N,A} + f(t_2) - f(t_1) \qquad (2)$$

Where:

$\varepsilon_{T,M,E,S,X,R,N,A}$ difference between the errors δ in the indications difference of the object and the standard;
$f(t_2)$ balance drift in the object weighing at $t_2$;
$f(t_1)$ balance drift in the object weighing at $t_1$.

While the effects due to systematic errors would be constant and with null true value [4], the difference between the drift in weighings could be non-zero. In fact, considering that the balance indications, even in unloaded condition, vary due to environmental effects and the heating of the balance in different times, the drift value would not be the same. With the purpose of eliminating the drift effect on the indication differences, several weighing cycles types can be applied [5]. At Inmetro's Mass Laboratory weighing cycles ABBA and ABABAB are routinely used, both are effective when the drift is linear.

The condition of equal time intervals between weighings applies to both weighing cycles. Typical results obtained in ultramicro comparator, with the application weighing series (cycles) ABABAB and ABBA are shown respectively in figures 1 and 2. The assumption of linear drift in equal time intervals might not be true for any type of curve drift.

Thus, in order to evaluate the efficacy in reducing or eliminating drift of the weighing in ABBA and ABABAB sequences, we used simulated drift curves with linear, quadratic, cubic, sine shapes with and without random term. The indication differences being determined for each weighing series in accordance with equations (3) and (4).

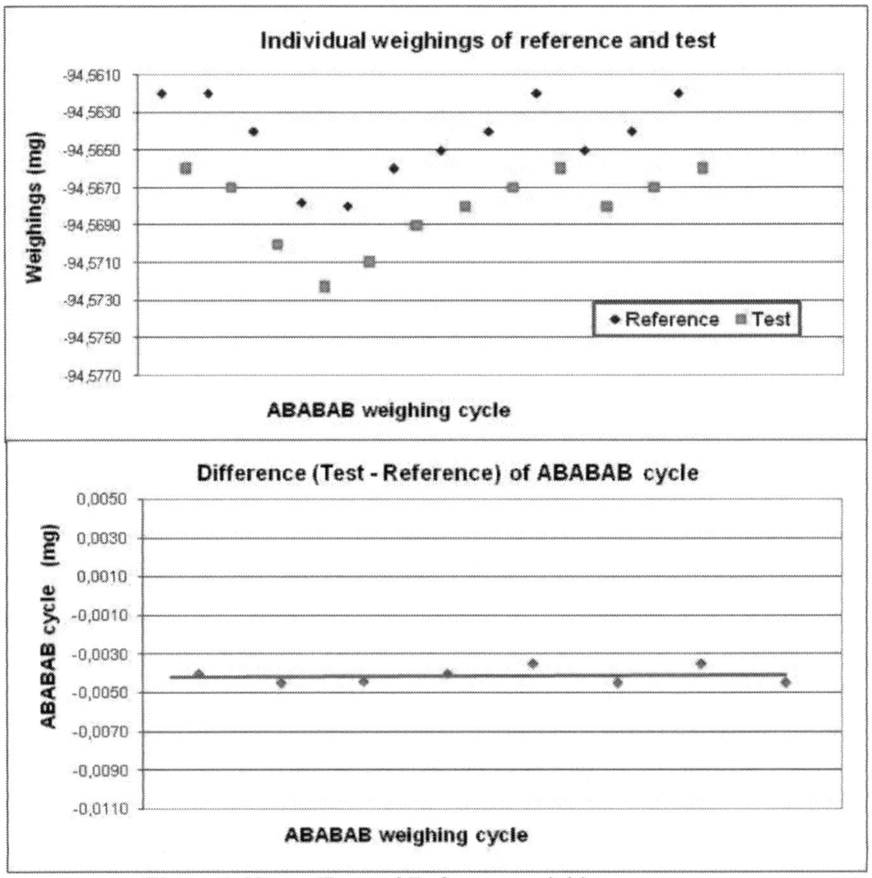


**Figure 1.** Upper: Test and Reference weighings.
Lower: drift corrected differences B – A for weighing cycles ABABAB

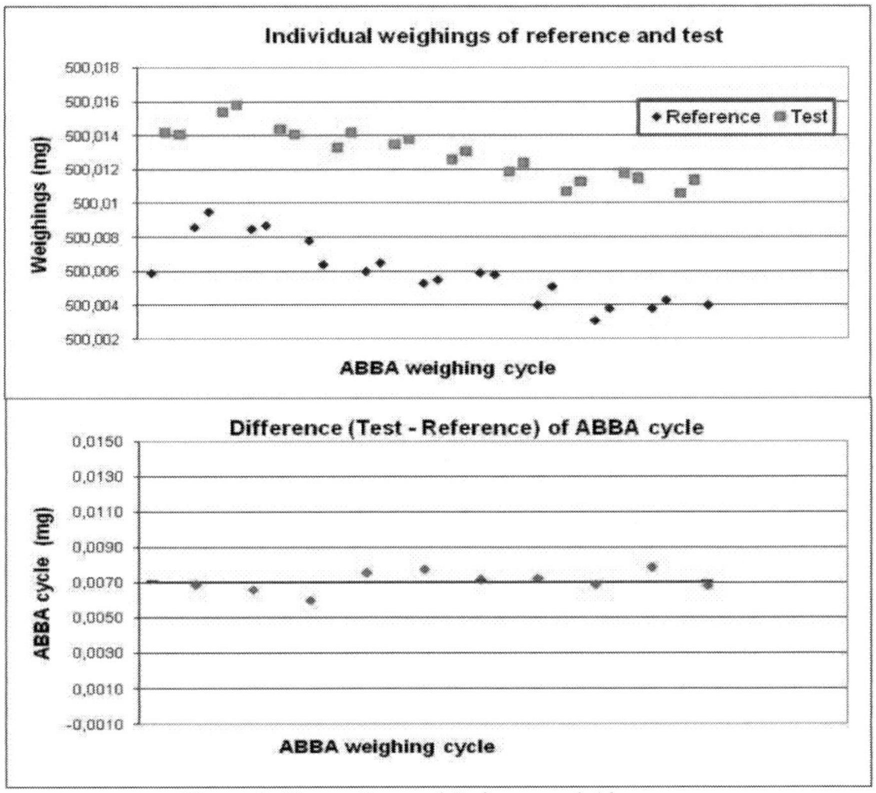


**Figure 2.** Upper: Test and Refernce weighings.
Lower: drift corrected differences B – A for weighing cycles ABBA

$$\Delta I_{\text{ABBA}} = \frac{(I_{B2} - I_{A2}) + (I_{B1} - I_{A1})}{2} \tag{3}$$

$$\Delta I_{\text{ABABAB}} = \begin{cases} I_{B1} - \dfrac{(I_{A1} + I_{A2})}{2} \; para \; o \; \text{ABA} \\ \dfrac{(I_{B2} + I_{B3})}{2} - I_{A3} \; para \; o \; \text{BAB} \end{cases} \tag{4}$$

Where:

$I_{Bi}$ balance indication for Test standard weighing $i$;
$I_{Ai}$ balance indication for Reference standard weighing $i$;
$i$ weighing order, corresponds to a constant time interval.

## 3. Drift simulation method

Two mass standards, A and B, were used to perform the zero drift simulation. The mass difference was set as being 0.2 (in mass arbitrary units). The reference standard mass value, $M_A$ , was assigned equal to zero. The test standard, B, has been assigned the mass value, $M_B$, equal to $M_A$ plus the amount of 0.2 mass units. Thus when there exists an exact cancellation of the balance drift, then the comparative weighing series will result in the mass difference value equal to 0.2.

The simulations are divided into two groups. In the first, simulated drift curves are continuous functions [6], where the dependent variable is the balance indication and the independent variable is time, the latter represented by the ordered number of the respective weighing. Regarding that each weighing takes place at equal time intervals. In the second group it was added a random term *A(t)* to the drift functions.

The random term has a rectangular probability distribution of width 1, which can be modulated by a multiplier factor. From these drift curves the weighing points are obtained in each of the ABBA and ABABAB weighing schemes for direct comparison of simulated results. In order to study the drift correction of random nature only, simulations were carried out with null drift, just acting the random term. A rectangular distribution and a Gaussian distribution for a series of up to one million repetitions were simulated.

### *3.1 Drift functions*

To simulate in a more comprehensive way the types of drift that may arise during the mass comparison process we used the drift functions shown in table 1, and the graphs of these functions are shown in figures 3 and 4.

**Table 1.** Drift equations

| Function | Drift equation | Parameters |
|---|---|---|
| Linear | $I = K \cdot N + C + M_{A,B} + A(t)$ | $K = 1.6 \quad C = -1.6$ |
| Quadratic | $I = K \cdot N^2 + C + M_{A,B} + A(t)$ | $K = 0.06 \quad C = -0.06$ |
| Cubic | $I = K \cdot N^3 + C + M_{A,B} + A(t)$ | $K = 0.002 \quad C = -0.002$ |
| Sine | $I = K \cdot sin(\pi \cdot C \cdot (N - 1)) + M_{A,B} + A(t)$ | $K = 30 \quad C = 0.05$ |

Where,

$I$ balance indication;
$K$ adjust coefficient;
$C$ adjust constant;
$A(t)$ random variable with adjustable amplitude;
$N$ weighing order (time of weighing);
$M_{A,B}$ mass value. Zero mass units for standard $M_A$ and 0.2 mass units for object mass $M_B$.

The simulations were performed in an Excel spreadsheet. As these are generic functions, the factors *K*, *C* and *M* were chosen to reproduce behaviors close to those observed in real measurements. But even with the simulation including other factors relative behaviors remain stable. Only for values *A(t)* greater than the mass difference (B – A) the drift functions are outweighed by the random term making equivalent the ABBA and ABABAB procedures.

In figure 3 it can be seen that the maximum drift range was chosen to approximately 40 mass units. Even in this drift range both weighing schemes can reduce the zero drift effects of the mass comparator. In figure 4 are shown the same functions as in figure 3, but with the addition of a random term (in figure 4 the oscillations due to random factor were amplified by a factor 5 in order to be more easily seen).

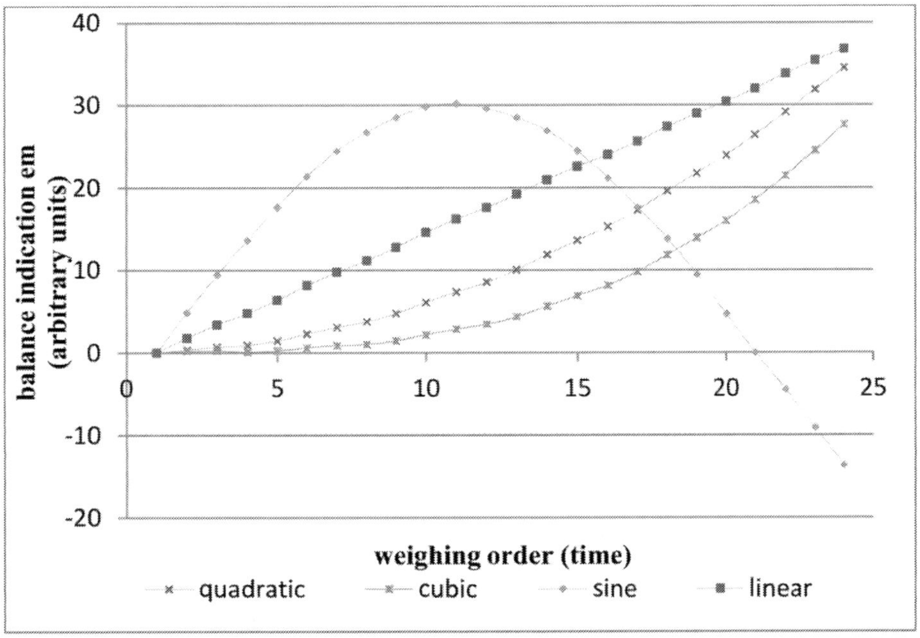


**Figure 3.** Drift functions

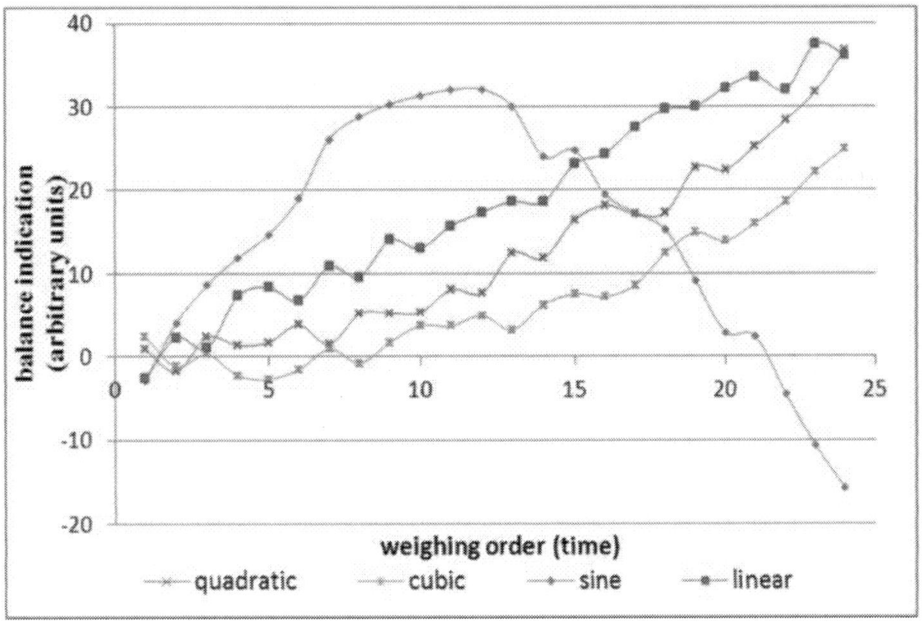


**Figure 4.** Drift functions with a random term

## 4. Results

The weighing simulation was performed for 24 virtual weighings which correspond to a typical measurement of $E_1$ accuracy class standards. This number corresponds to 6 weighing cycles in ABBA scheme and to 8 cycles ABA and BAB. The simulations results of the weighing sequences are shown in table 2.

Journal of Physics: Conference Series **648** (2015) 012025 doi:10.1088/1742-6596/648/1/012025

**Table 2.** Simulations results

| Drift | $\Delta m_{ABBA}$ | $\Delta m_{ABABAB}$ | $\sigma_{max}$ | $\Delta m_{ref}$ |
|---|---|---|---|---|
| Linear | 0.200 | 0.200 | 0.000 | 0.200 |
| Quadratic | 0.008 | 0.200 | 0.064 | 0.200 |
| Cubic | 0.050 | 0.209 | 0.091 | 0.200 |
| Sine | 0.567 | 0.211 | 0.339 | 0.200 |

The $\Delta m$ values correspond to the mass difference of each weighting scheme, $\Delta m_{ref}$ is the theoretical reference value and $\sigma_{max}$ is the maximum standard deviation obtained from two weighing schemes. Results indicated that both methods are suitable to remove the purely linear drift. The ABABAB scheme is still accurate for the elimination of quadratic drift. In the case of cubic and sine drifts both schemes do not completely eliminate the drift effects but ABABAB scheme was more effective in reducing the drift effects.

These results remain the same even if the random term is included, since the variation is typically less than the mass difference to be measured. Thus for a mass difference of 0.2 arbitrary mass units, a random variation in the interval $[-0.1, 0.1]$ was chosen and the obtained results are shown in table 3.

**Table 3.** Simulations results with random term

| Drift + $A(t)$ | $\Delta m_{ABBA}$ | $\Delta m_{ABABAB}$ | $\sigma_{max}$ | $\Delta m_{ref}$ |
|---|---|---|---|---|
| Linear | 0.214 | 0.191 | 0.008 | 0.200 |
| Quadratic | 0.069 | 0.199 | 0.063 | 0.200 |
| Cubic | 0.019 | 0.182 | 0.091 | 0.200 |
| Sine | 0.602 | 0.199 | 0.339 | 0.200 |

In a linear drift with random term acting simultaneously there is an important difference between the analyzed weighing schemes. Again the ABABAB scheme was superior to ABBA scheme for reducing the drift effects, even considering the maximum standard deviation $\sigma_{max}$, the mass differences values quadratic drift, cubic and sine, obtained by ABBA scheme does not reach the 0.2 difference value. Even the ABBA scheme providing lower standard deviations than the ABABAB scheme, the final results were out of the theoretical value to be obtained, 0.200 mass unit.

*4.1 Isolated random term effect*

The simulations with random term and without drift, have inconclusive results for 24 weighing series.

**Table 4.** Random drift effect (arbitrary units)

| Distribution | | ABBA | ABABAB | Average |
|---|---|---|---|---|
| Retangular | $\Delta m$ | -0.0001 | -0.0002 | 0.0001 |
| | $\sigma$ | 0.0482 | 0.0510 | 0.0240 |
| Gaussian | $\Delta m$ | 0.0108 | 0.0052 | -0.0019 |
| | $\sigma$ | 0.2142 | 0.2151 | 0.1127 |

Thus, series of $1x10^6$ repetitions of 24 weighings were performed with a rectangular distribution and a Gaussian distribution with adjustable width (range). In both cases the results indicate that there is no difference in applying the ABBA and ABABAB schemes because indication differences always remain one to two orders of magnitude smaller than the standard deviations involved in the simulation.

As an example, some results of simulations with $1x10^6$ random repetitions are shown in table 4, where the results corresponds to width range equal to 1.

*4.2 Experimental data analysis*

For a comparative analysis with real measurements three sets of display readings were performed using mass comparator Mettler-Toledo AT1006. The reading of the comparator indications are given without the use of mass standards (empty plate), with the mass comparator only performing weighing operation (with internal weights) and under the influence of the equipment drift and the environment conditions.

The series of 180 weighings are shown in figures 5, 6 and 7. In the first sequence, it was obtained an average value of □0.0002 mg from the 180 weighing indications. By applying the ABBA weighing scheme the result of the elimination of drift is 0.0000 mg, meanwhile the ABABAB scheme resulted in 0.0002 mg with their respective standard deviations of 0.00035 and 0.00044 mg.

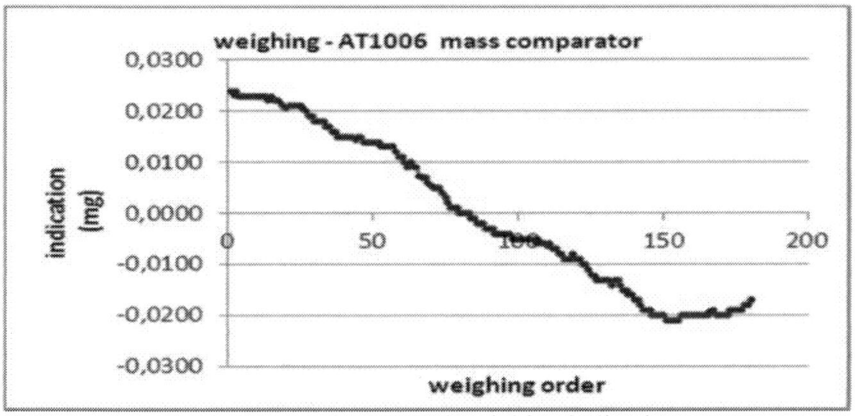


**Figure 5.** AT1006 weighing indications

The balance resolution is 1 microgram, but by performing the integration of the last digit for 10 seconds it gets an effective resolution of 0.1 micrograms.

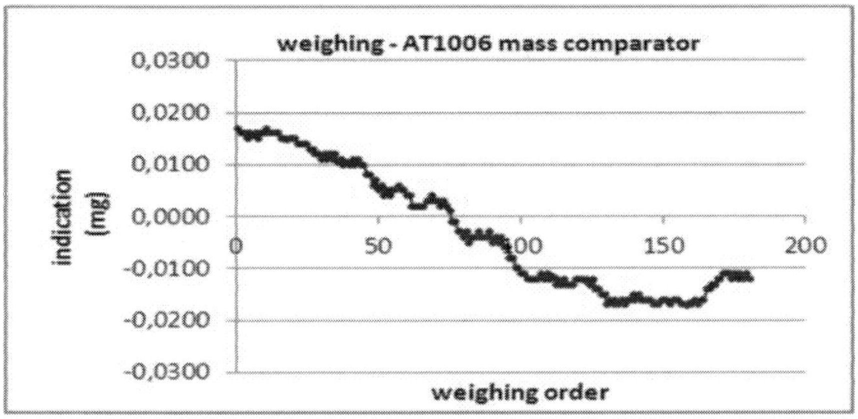


**Figure 6.** AT1006 weighing indications

In the second sequence, figure 6, the average measures resulted in 0.0029 mg. In ABBA scheme it was obtained 0.0001 mg and in ABABAB scheme 0.0007 mg, with standard deviations, respectively, 0.00039 and 0.00039 mg. In the third sequence of 180 weighings, figure 7, it was obtained an average of 0.0014 mg, the ABBA scheme results 0.0001 mg and 0.0007 mg in ABABAB, with standard deviations respectively 0.00035 and 0.00037 mg.

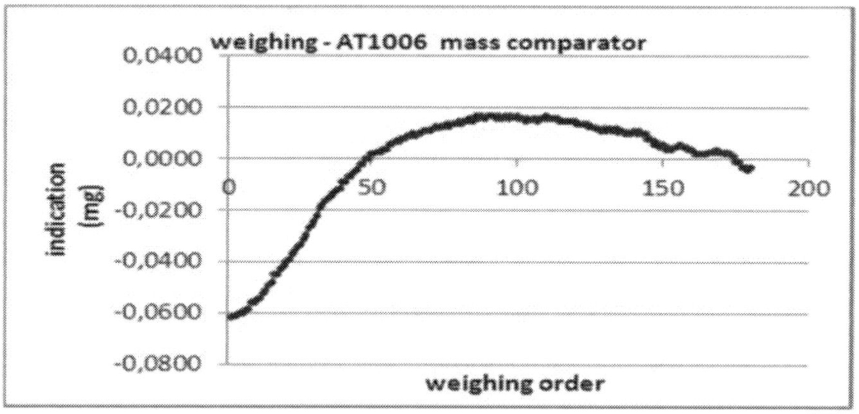


**Figure 7.** AT1006 weighing indications

In this case, unlike the simulation, do not have an exact value to be achieved. Because it is a zero drift measurement it is expected that, in a sufficiently large time interval (infinite measurements), a null average value will be obtained, but being an empty plate measurement thus zero is the value which is to be expected as mass difference. In a short time, the average value of the measurements can not be taken as the expected value. The drift may occur in such a way that the average value does not correspond to the null value as would be expected in a system with random zero drift behavior. ABBA and ABABAB schemes are designed in order to, under the specified conditions, reduce the drift effects. So by weighing without a test object on the balance weighing plate, both weighing schemes should get null value for the mass difference.

In the first weighing, both weighing schemes were equivalent because the null value obtained was within the standard deviation. In the second and third sequences the ABABAB remained more than one standard deviation away from one standard deviation, while the ABBA scheme achieved a result within the standard deviation.

## 5. Conclusion

According to the results, both cycles eliminated the drifts of linear nature, when a random nature variation is added to the linear behavior of both weighing schemes are still equivalent in treatment. In the drifts of quadratic nature the elimination of effects is achieved only in the ABABAB cycle.

In cubic and sinusoidal drifts the ABABAB advantage occurs at the same rate, with or without the addition of the random term, provided that the last one have amplitude lesser than the mass differences. Despite the ABABAB cycle has a disadvantage for providing results with slightly larger standard deviations [7, 8, 9] than the ABBA cycle, this disadvantage is offset by better results from differences in indication for the studied theoretical cases.

The theoretical identification of this advantage in mass difference results, in case of real weighings, given the small number of comparison cycles, the drift magnitude and its variations are more complex functions than the ones studied here, also in tests with real weighings (just a few cycles) the differences of the achieved results in both weighing schemes are very close to each other, making it inconclusive the assessment of the effectiveness of both treatments for real measurements.

The random drift analysis shows that there is no advantage for any of the studied comparative weighing schemes. In tests with one million simulated weighing results both approaches eliminated the effects of the random nature drift. When the effects of random term are amplified, the behavior between the ABBA and ABABAB schemes approaches the purely random behavior, that is because the random fluctuations dominate the character of the chosen drift function.

In this work it was identified differences in the zero drift elimination characteristics and the calculated standard deviations between comparison weighing schemes, ABBA and ABABAB, but there is still room to study these schemes with more complex functions, function combinations and real measurements.

## 6. References

[1] OIML D28 2004 *Conventional value of the result of weighing in air* Edition 2004 (E).

[2] OIML R111-1 2004 *Weights of classes* $E_1$, $E_2$, $F_1$, $F_2$, $M_1$, $M_{1-2}$, $M_2$, $M_{2-3}$ *and* $M_3$, *Part 1: Metrological and technical requirements*. Edition 2004 (E).

[3] NIT-Lamas-010 2013 *Determinação da Incerteza na Calibração de Balanças*, rev.02, Inmetro/Dimci/Dimec/Lamas.

[4] VIM 2012 *Vocabulário Internacional de Metrologia: conceitos fundamentais e gerais de termos associados*, Inmetro.

[5] Sutton C M and Clarkson M T 1993/1994 *A General Approach to Comparisons in the Presence of Drift. Metrologia*, vol. **30**, pp. 487-493.

[6] Beatrici A, Cacais F L, Quintão D N, Rebelo Jr A, Loayza V M 2013 *Eficácia dos Métodos de Comparação de Massa no Tratamento da Deriva em Balanças Comparadoras*. 7º Congresso Brasileiro de Metrologia, Ouro Preto – MG.

[7] Gläser M 2000 *Cycles of comparison measurements, uncertainties and efficiencies* Meas. Sci. Technol., vol. 11, pp. **20-24**.

[8] H Erik Swanson and Stephan Schlamminger 2010 *Removal of zero-point drift from AB data and the statistical cost* Meas. Sci. Technol. **21**.

[9] Gläser M 2003 *Uncertainty and efficiency of correlated measurement cycles with periodically varying patterns* Meas. Sci. Technol. **14**.

**Acknowledgments**

Thanks to Pronametro scholarship program for financial support.